Michael Quinten

Practical Determination of Optical Constants from Spectral Measurements

The author

Dr. rer. nat. habil. Michael Quinten works as Head of R&D Sensors at FRT GmbH in Bergisch Gladbach, Germany. During his academic period from 1983 to 2000 he obtained his diploma degree and Ph. D. in physics at the University of Saarland, Saarbruecken, Germany, and joined the Technical University RWTH Aachen for habilitation. In 2001 he joined the ETA-Optik GmbH, Germany, where he first worked in research and development of integrated optics components and later became product manager in the Colour and Coatings Division. In 2007 he moved to FRT GmbH where he is responsible for the optical sensor technology division.

Based on comprehensive knowledge in solid state physics, nanomaterials, and optical sensor technology, he authored more than 50 scientific publications with several topics and three books.

M. Quinten
Optische Schichtdickenmessung mit miniaturisierten Spektrometern
BoD - Books on Demand, Norderstedt (2015)
ISBN 978-3-7347-8386-9

M. Quinten
A Practical Guide to Optical Metrology for Thin Films
Wiley-VCH, Weinheim (2012)
ISBN 978-3-527-41167-4

M. Quinten
Optical Properties of Nanoparticle Systems - Mie and Beyond
Wiley-VCH, Weinheim (2011)
ISBN 978-3-527-41043-9.

Michael Quinten

Practical Determination of Optical Constants from Spectral Measurements

First Edition

BoD - Books on Demand, Norderstedt, 2018

Bibliografische Information der Deutschen Nationalbibliothek:
Die Deutsche Nationalbibliothek verzeichnet diese Publikation in der Deutschen Nationalbibliografie; detaillierte bibliografische Daten sind im Internet über http://dnb.dnb.de abrufbar.

© 2018 Michael Quinten

Herstellung und Verlag: BoD – Books on Demand, Norderstedt

ISBN: 978-3-7460-2486-8

Contents

1 Introductory Remarks 1
2 Physical Basics 5
3 The Modeling of Optical Constants 13
    3.1 Physical Models of the Dielectric Function 13
        3.1.1 The Harmonic Oscillator Model 13
        3.1.2 The Drude Dielectric Function 18
        3.1.3 Extensions of the Harmonic Oscillator Model 21
        3.1.4 Kramers Kronig Relations 30
    3.2 Empirical Models for the Refractive Index 32
    3.3 Effective Medium Approaches 36
4 The Practical Determination of Optical Constants 41
    4.1 Spectral Reflectance Measurement 46
    4.2 Spectral Ellipsometric Measurement 56
5 Guidelines to the Practical Determination of Optical Constants . 61
    5.1 How to Find Proper Models 62
    5.2 How to Adjust Initial Values for the Fit 80
    5.3 How to Verify the Results 83
6 Appendices 85
    6.1 Appendix A: Numerics With Complex Numbers 85
    6.2 Appendix B: Levenberg-Marquardt Algorithm 89
7 References 93

1 Introductory Remarks

Optical constants are specific properties of condensed matter that allow to describe in a simple way the interaction of light or other electromagnetic radiation with matter. There is always a requirement for optical constants values for estimating colour, reflection, internal total reflection, refraction, scattering, phase shifting, multilayer properties, or thin film thickness.

In general, we can distinguish between the dielectric constant ε and the refractive index n and the extinction (or absorption) index κ as optical constants, although they are strongly related to each other. The generally complex *dielectric function* $\varepsilon(\omega) = \varepsilon_1(\omega)+i\cdot\varepsilon_2(\omega)$ with the imaginary unit $i=\sqrt{-1}$ is the physically relevant quantity for describing the interaction of light with matter, while for discussing propagation of light in and through media the generally complex *refractive index* $n+i\cdot\kappa$ is the more proper quantity.

The diversity of condensed matter with insulators, oxides, semiconductors, compound semiconductors, metals, and alloys calls for a huge amount on various optical constants even in different spectral regions. Therefore, there exist numerous works on the determination of optical constants with either tabulated data, graphical representation, or a parametrization of the data for a certain model.

The *Handbook of Optical Constants of Solids* [1], edited by Edward D. Palik, affords the most comprehensive database of the refractive index and absorption index of technically important and scientifically interesting dielectrics, semiconductors, and metals in three volumes. The online version of the *Handbook of Optical Constants of Solids* (a set of even five volumes) can be found on ScienceDirect.com. Beyond that, the database is supplemented by tutorial chapters covering the basics of dielectric theory and reviews of experimental techniques for each wavelength region and material. Although these tutorial chapters contain a huge amount of information, not all described experimental techniques are up-to-date or recently developed techniques are not contained. Another comprehensive sources of tabulated data are the *Handbook of Optics II* [2], the *CRC Handbook of Chemistry and Physics* [3], and particularly for semiconductors the book *Optical Constants of Crystalline and Amorphous Semiconductors: Numerical Data and Graphical Information* [4] by Sadao Adacho. A special reference for particularly thin films is the book *The Optical Constants of Bulk Materials and Films* by L. Ward [5]. It covers the theoretical background, experimental techniques, and results for a wide range of materials of thin films. Beyond that optical constants are published in further books and in numerous articles in several journals. Many of the data in these references are also available from several online databases, the most prominent being http://refractiveindex.info and www.luxpop.com. Also some suppliers of measurement equipment provide databases, e.g. Filmetrics Inc. at http://www.filmetrics.com/refractive-index-database.

With regard to this huge amount of information on optical constants and methods how to determine optical constants the question may arise why this booklet is necessary at all? The answer is that in many modern applications where optical constants are needed often there is a lack of knowledge about such data. They are either not contained in any collection or simple methods for a pretty fast determination are missing. This booklet has the aim to shorten this gap by introducing briefly in two simply accomplishable measuring methods -spectral reflectometry and spectral ellipsometry- and in available models for the parametrization of optical constants for determination of n and κ from the spectral measurements.

It begins with some few physical basics of light and the interaction of light with matter as well as information on the relationships among (complex) dielectric function and (complex) refractive index. Then, we introduce in physical models for the dielectric function and in empirical models mainly developed for the refractive index. The latter originate from the demand to have a model for the dispersion of the refractive index -this is the wavelength dependence of the refractive index- of those materials that are used in optical instruments like microscopes and telescopes. Also the specialty of the optical constants of composites of at least two nonmixable components is considered. The next chapter is dedicated to methods and techniques for the determination of optical constants. The focus is set to spectroscopic reflectometry and spectroscopic ellipsometry, as they can easily be applied mainly in the most interesting spectral range from the near ultraviolet to the near infrared. They always result in spectra of characteristic quantities that are determined by the optical constants. The task is then to retrieve the optical constants from these spectra. This task can only be solved using complex algorithms together with a parametrization of the optical constants. So we give finally some guidelines for the practical determination of optical constants from spectral reflectometry and ellipsometry in the last chapter.

To write this booklet required reading and valuating of many monographs and a still larger number of publications on this subject. The total amount of published work is too immense to consider them all here. Therefore, I hope to have included the most relevant up to date, and apologize for all the contributions not considered here.

Last but not least, I want to thank my family for their support and patience with me during writing this booklet.

Aldenhoven, 05.02.2018   Michael Quinten

2 Physical Basics

Light or in general electromagnetic radiation can be described with the model of a wave that propagates with velocity of light c in a straight line in the vacuum. A wave is in general a process that is periodic in space *and* time, i.e. W = W(r, t). That means there is a *time period* T after which the wave W looks the same as at the timepoint t, i.e. W(r, t+T) = W(r, t). The reciprocal value 1/T defines the *frequency* ν of the wave. Similar holds for the three-dimensional space. There is a periodicity **R** after that the wave appears the same as at a certain point **r**, i.e. W(**r**+**R**, t) = W(**r**, t). The modulus |**R**| of this periodicity vector is the *wavelength* λ, defining the distance between two wave peaks.

The entity of electromagnetic radiation is illustrated in Fig. 2.1. The spectral range accessible to the human eye is called *light* and is rather small. It is indicated in Fig. 2.1 by the coloured region in the mid of the graphics.

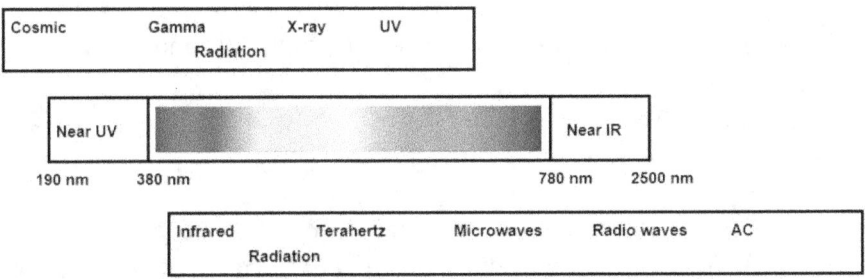

Fig. 2.1 The complete spectral range of electromagnetic radiation. The coloured region from 380 nm to 780 nm is the visible spectral range

The spectral ranges are usually specified in different units, depending on the application. Specifications in use are:

electron volt (eV): for the energy E of a radiation quantum, commonly used for visible light and radiation of higher frequency (UV, X-ray, ...),
wavelength (nm): for the wavelength λ of the radiation, commonly used for UV, visible light, and near infrared,
wavelength (µm): for the wavelength λ of the radiation, commonly used for near and mid infrared, sometimes also for visible light,
wavenumber (cm^{-1}): for the reciprocal wavelength $\tilde{\nu} = 1/\lambda$, commonly used for near to far infrared, sometimes also for visible light.

Often *angular frequency* ω (1/s) is used instead of the frequency ν (Hz). They are related to each other via

$$\omega = 2\pi \cdot \nu. \tag{2.1}$$

All the quantities above can be converted from one to another. For this the following relations must be known:

$$\lambda \cdot \nu = c \tag{2.2}$$

for wavelength λ, frequency ν and velocity of light c, and

$$E = h\nu = \hbar\omega = \frac{h \cdot c}{\lambda} \tag{2.3}$$

for the energy of a radiation quantum of frequency ν respectively wavelength λ. Additionally the values of the natural constants Planck's constant $h = 6.62606896 \cdot 10^{-34}$ Js, $\hbar = h/(2\pi)$, velocity of light in vacuum $c = 2.99792458 \cdot 10^8$ m/s, and elementary charge $e_0 = 1.60217648 \cdot 10^{-19}$ As (according to PTB and NIST), and the value of $\pi = 3.14159265359$ must be known. Then

$$E\,[eV] = \frac{1.23984187542}{\lambda\,[\mu m]} = \frac{1239.84187542}{\lambda\,[nm]}, \tag{2.4}$$

$$\omega[s^{-1}] = \frac{1.88365156731}{\lambda\,[\mu m]} = \frac{1883.65156731}{\lambda\,[nm]}, \tag{2.5}$$

$$\tilde{\nu}\,[cm^{-1}] = \frac{10000}{\lambda\,[\mu m]} = \frac{10^7}{\lambda\,[nm]}. \tag{2.6}$$

When talking about electromagnetic waves, it is not mandatory, but often to find that a harmonic wave in time and space is given by

$$W(\mathbf{r}, t) = W_0 \cdot \exp(i \cdot (\mathbf{k}\mathbf{r} - \omega t + \phi)) \tag{2.7}$$

with **k** being the *wavevector* that describes the propagation direction of the wave. Its norm is $k = 2\pi/\lambda$. The quantity ϕ is an arbitrary constant. The complete expression **k·r** - ω·t + ϕ is the *phase* of the wave. For an *electromagnetic* wave we have to consider an *electric field* **E**(r,t) and a *magnetic field* **H**(r,t) that can be described by Eq. (2.7) and must fulfill Maxwell's equations. (For Maxwell's equations see textbooks on classical electrodynamics from e.g., Born and Wolf [6], Stratton [7] or Jackson [8].) This description with complex numbers enormously simplifies the calculation of electromagnetic fields as well as further related quantities. For a brief introduction in the numerics with complex numbers see Appendix A.

As in vacuum by definition does exist nothing, no pertubation of the propagation of light by interaction with matter can occur. However, already one single hydrogen atom can destroy this harmony. The reasons are the force fields of the electromagnetic radiation: the electric field **E** and the magnetic field **H**. On the other hand, an atom consists of heavy electrically neutral (neutron) or positively charged (proton) tiny components that form the core of the atom, surrounded by negatively charged electrons with a mass of about 1/2000 of the proton mass that orbit the core like planets the sun (*Bohr's atom model*). There are as much electrons as protons to make the atom electrically neutral.

An electrical charge q in an electric field experiences a force F_e proportional to the electric field **E** and the size of the charge. In a magnetic field **H** the charge experiences a second force F_m also proportional to the magnetic field and the size of the charge, but perpendicular to the direction of movement, if the charge is moving. The magnetic force is about 400 times smaller than the electric force because the magnetic field of an electromagnetic wave is about 400 times smaller than the electric field.

Caused by these forces the center of gravity of the positively charged protons separates from the center of gravity of the negatively charged electrons, as both charges move in opposite direction. Actually, the protons are as heavy as under normal conditions of light ($|E|$ < 100 V/m) mainly the electrons get displaced. Then, the atom becomes an *electrical dipole* **p** = q·**r** (two opposite charges of same size with distance **r** between the centers of gravity). As the fields of an electromagnetic wave periodically change in time also the dipole changes periodically. That means that within a period of T the dipole changes from plus-minus in the first half period to minus-plus in the second half period. Hence, (mainly) the electrons experience a permanent acceleration by the electric and magnetic forces. The energy contained in this acceleration (from work = force x path, W = q·**E**·**r** = **p**·**E**) would destroy the atom within a short time for which reason the dipole reemits energy as an electromagnetic wave.

When going on from the single atom to molecules, clusters, nanoparticles up to solid regular assemblies of atoms (crystals) or fluids, the description of the interaction of electromagnetic waves with matter becomes more and more complex. With increasing number of atoms in the assembly specific electron distributions form around the atoms. In a solid state body they are known as *band structure*. Then, also the scope of interactions of light with matter increases. Now, one must define models that consider the specific electron distribution, the acceleration of the electrons in electric and magnetic fields, and exchange and interaction effects among the electrons to describe the interaction of light with matter. The most precise description is only possible using quantum mechanics, but also from classical mechanics very well suited descriptions are available. The result of all these examinations is the *complex dielectric function* $\varepsilon(\omega) = \varepsilon_1(\omega)+i\cdot\varepsilon_2(\omega)$. With the dielectric function we introduce a quantity that gives us information on magnitude and type of interaction of the electromagnetic wave with matter. We will be concerned with it in the following chapters again and again.

The classical interpretation of ε_1 and ε_2 is as follows. For the real part ε_1 we consider first an empty capacitor in Fig. 2.2. The capacitance C is given as

$$C = \varepsilon_0 \cdot \frac{A}{d} \tag{2.8}$$

and is limited by the area A of the capacitor plates and the distance d between the plates.
If we put a dielectric material in between the plates the capacitance increases by a factor ε

$$C = \varepsilon \cdot \varepsilon_0 \cdot \frac{A}{d}. \tag{2.9}$$

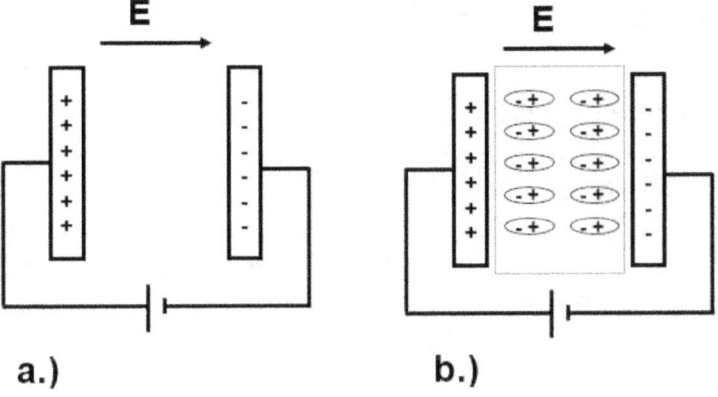

a.)   b.)

Fig. 2.2 a.) Empty capacitor, b.) Capacitor filled with a dielectric material

The reason is that the electric field between the two plates induces a dipole at each atom. They sum up to a net polarizability **P** that adds to the electric field **E** between the plates. The factor ε here corresponds to the real part ε_1 of the dielectric function. That means, the real part of the dielectric function contains information on the polarizability of matter.

For interpretation of the imaginary part ε_2 we must consider another process, namely the *absorption*. Passing through an absorbing medium of thickness d, we can observe that the intensity I_0 gets attenuated. This is illustrated in Fig. 2.3.

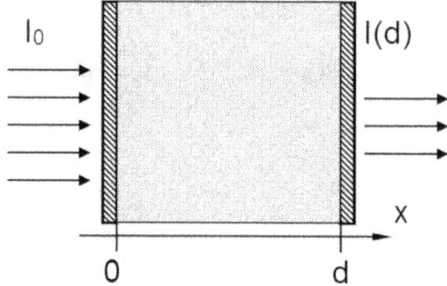

Fig. 2.3 Light passing through an absorbing medium

The transmitted intensity I(d) is

$$I(d) = I_0 \cdot \exp(-\alpha \cdot d) \qquad (2.10)$$

This is *Lambert-Beer's law*. The factor α in the exponential function is the *absorption coefficient*. It is directly connected to the imaginary part of the dielectric function ε_2 via

$$\alpha(\omega) = 2 \cdot \frac{\omega}{c} \varepsilon_2(\omega) \qquad (2.11)$$

If the wave gets attenuated it loses energy which is released to the matter. There, molecular vibrations or lattice vibrations (phonons) are excited.

If one considers only the propagation of the wave in matter one recognizes that the propagation velocity gets reduced caused by the interaction with the matter. Can the matter optically be described by a *refractive index* n(ω), the propagation velocity reduces to $\frac{c}{n(\omega)}$. Passing through an absorbing medium with *complex refractive index* $\hat{n}(\omega) = n(\omega) + i \cdot \kappa(\omega)$ also the magnitudes of the electric and magnetic field get attenuated whereat the *extinction coefficient* or *absorption index* κ(ω) plays an important role.

Dielectric function and refractive index are connected by *Maxwell's relation*

$$n + i\kappa = \sqrt{\varepsilon_1 + i\varepsilon_2} \,. \tag{2.12}$$

This relation can be rewritten either for the dielectric function as

$$\varepsilon_1 = n^2 - \kappa^2, \tag{2.13}$$

$$\varepsilon_2 = 2 \cdot n \cdot \kappa, \tag{2.14}$$

or for the refractive index as

$$n = \sqrt{\frac{\varepsilon_1}{2} + \frac{1}{2}\sqrt{\varepsilon_1^2 + \varepsilon_2^2}}, \tag{2.15}$$

$$\kappa = \sqrt{-\frac{\varepsilon_1}{2} + \frac{1}{2}\sqrt{\varepsilon_1^2 + \varepsilon_2^2}}. \tag{2.16}$$

Now it is about time to point to a particularity of the dielectric function: the *anisotropy* of the dielectric function. This pecularity is not obvious. However, when going into the details of the electronic structure of a crystal, we can find a dependence of the electron distribution on the crystalline structure. The distribution of electrons is not equal in each crystalline direction. This is relevant mainly for tetragonal and orthorhombic crystal structures. In consequence, the dielectric function is also different in different directions because the induced dipoles differ for each crystalline direction. This anisotropy results in an *ordinary ray* that obeys the common rules in optics (reflection law, Snell's law) and in at least one *extraordinary ray* with specific description and rules. The dielectric function is now a tensor of second rank.

3 The Modeling of Optical Constants

Optical constants are helpful for the description of the interaction of electromagnetic radiation with matter. Hence, they can also be represented by an according model. In this chapter we introduce in physical models and the classical description of $\varepsilon(\omega)$ from a macroscopic view. For a quantum mechanical derivation and discussion of the corresponding results we refer e.g. to [9] and references given therein or to relevant textbooks in solid state physics, e.g. [10]. The classical description results in the *harmonic oscillator model* that is of much greater interest in practical applications than the quantum mechanically derived dielectric function. Basing on this harmonic oscillator model we discuss the *Drude dielectric function* and some extensions of the harmonic oscillator. With the *Kramers Kronig relations* relationships between real and imaginary part of the optical constants are established that are helpful particular in determination of optical constants.

Although the physical models are always applicable there exist some empirical models for especially the refractive index which are still in use up to date. They are briefly reviewed in Sect. 3.2 of this chapter. A specialty is finally discussed in Sect. 3.3 for microscopically inhomogeneous materials, i.e. materials where two or more components of minute size are distributed in a homogenous material, forming a so-called *effective medium*.

3.1 Physical Models of the Dielectric Function

3.1.1 The Harmonic Oscillator Model

As already mentioned in Chapter 2 interaction of electromagnetic fields with matter is dominated by the forces exerted by the incident electric (and magnetic) field on the electric charges in the matter. At high frequencies the electric field **E** inside the body of condensed matter usually displaces the electrons in the atoms of condensed matter while the ions are too inert as to follow the electric field with the same frequency. Thereby each atom becomes an electric dipole with dipole moment **p**.

The dipole moments add up to a macroscopic net polarizability **P** of the sample. The connection between **P** and **E** can be described by the general equation

$$\mathbf{P} = \varepsilon_0 \chi \mathbf{E} \tag{3.1}$$

defining the macroscopic *susceptibility* χ of the matter and ε_0 = 8.854·10^{-12} As/Vm is a natural constant. Physics enters this formal relation by interpreting χ.

Charges q_n that are displaced from their position of equilibrium, are retreated by forces which are proportional to the distance from the position of equilibrium, similar to a mechanical spring. In consequence, a charge q_n with mass m_n executes forced oscillations in a time-periodic electric field $\mathbf{E} = \mathbf{E}_0 \cdot \exp(-i \cdot \omega t)$. This is the *harmonic oscillator model*. It was developed by H. A. Lorentz at the beginning of the 20th century [11] and is also called *Lorentz oscillator*.

From the force balance we obtain the following differential equation for the displacement r_n:

$$\sum_n m_n \frac{\partial^2 r_n}{\partial t^2} + m_n \gamma_n \frac{\partial r_n}{\partial t} + D_n r_n = \sum_n q_n E(t) . \tag{3.2}$$

The second term with γ_n accounts for the perturbation of the movement by interactions with other charges and lattice defects. It corresponds to a mechanical friction in a classical sense. The solution of this differential equation is given as

$$r_n = \frac{q_n}{m_n} \frac{E_0}{(\omega_n^2 - \omega^2 - i\omega \gamma_n)} \tag{3.3}$$

with the *resonance frequency* $\omega_n^2 = D_n/m_n$. If ω approaches ω_n the denominator becomes minimal and the displacement maximum. Without pertubation, i.e. $\gamma_n = 0$, the displacement becomes even infinite because the denominator approaches then zero (*resonance catastrophe*).
The macroscopic net polarizability **P** follows from all dipole moments $\mathbf{p}_n = q_n \mathbf{r}_n$ in the sample volume V via

$$\mathbf{P} = \frac{1}{V}\sum_n N_n \mathbf{p}_n = \frac{1}{V}\sum_n N_n q_n \mathbf{r}_n , \qquad (3.4)$$

from which we obtain the susceptibility χ_n for all N_n charges q_n as

$$\chi_n(\omega) = \frac{1}{V}\sum_n \frac{N_n q_n^2}{\varepsilon_0 m_n} \frac{1}{(\omega_n^2 - \omega^2 - i\omega\gamma_n)} . \qquad (3.5)$$

All contributions χ_n add up to a total susceptibility χ from which the *dielectric function* $\varepsilon(\omega) = \varepsilon_1(\omega) + i\cdot\varepsilon_2(\omega)$ follows as

$$\varepsilon(\omega) = 1 + \chi = 1 + \sum_n \frac{\omega_{Pn}^2}{(\omega_n^2 - \omega^2 - i\omega\gamma_n)} . \qquad (3.6)$$

The abbrevation

$$\omega_{Pn}^2 = \frac{N_n q_n^2}{V \varepsilon_0 m_n} \qquad (3.7)$$

is called *plasma frequency* of the n-th oscillator.
For small frequencies $\omega \to 0$ the real part $\varepsilon_1(0)$ of the dielectric function becomes constant

$$\varepsilon_1(0) = 1 + \sum_n \frac{\omega_{Pn}^2}{\omega_n^2} , \qquad (3.8)$$

while the imaginary part $\varepsilon_2(0)$ vanishes. $\varepsilon_1(0)$ represents the static dielectric constant of the material. The constant ratio $\dfrac{\omega_{Pn}^2}{\omega_n^2}$ defines a new quantity, the *oscillator strength* S_n of the n-th harmonic oscillator, so that ω_{Pn}^2 in (3.6) can be replaced by $S_n \omega_n^2$.

Fig. 3.1 exemplarily shows the dielectric function and the corresponding refractive index of a harmonic oscillator (Lorentz oscillator) with resonance frequency $3.5 \cdot 10^{15}$ s^{-1}, damping constant $3.5 \cdot 10^{14}$ s^{-1} and oscillator strength S = 1.

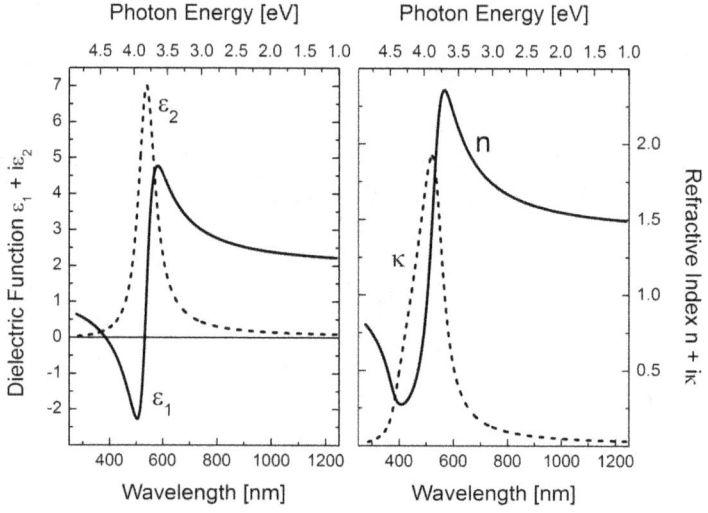

Fig. 3.1 Dielectric function and refractive index of a harmonic oscillator with oscillator strength S = 1, resonance frequency $3.5 \cdot 10^{15}$ s^{-1}, and damping constant $3.5 \cdot 10^{14}$ s^{-1}

At the resonance frequency the imaginary part exhibits a maximum and rapidly decreases to the right and the left of the resonance frequency. Far from the resonance frequency ε_2 vanishes. The real part decreases at high frequencies when approaching the resonance frequency and even becomes negative. In the vicinity of the resonance frequency the real part changes very rapidly to high positive values from which it continously decreases again when going to low frequencies. The decrease with decreasing frequency is called *normal dispersion* while in the region of rapid increase with decreasing frequency it is called *anomal dispersion*.

In general, the harmonic oscillator model can be applied in all cases where the force fields of the electromagnetic wave can induce electric dipoles by separating the center of gravity of positive charges and negative charges. Therefore, not only electronic excitations can be described with this model but also atomic excitations in bipolar crystals like NaCl or excitations of dipoles like in water (the biggest natural dipole). Then, the dielectric function is the sum over all the susceptibilities resulting from these excitations. As the ions in bipolar crystals are much heavier than electrons the corresponding atomic excitations lie at low frequencies in the far infrared where the negatively and positively charged ions are displaced in different directions. At frequencies in between all the excitations of dipoles due to a change of the orientation of the dipoles caused by the electric field can be found (see Fig. 3.2).

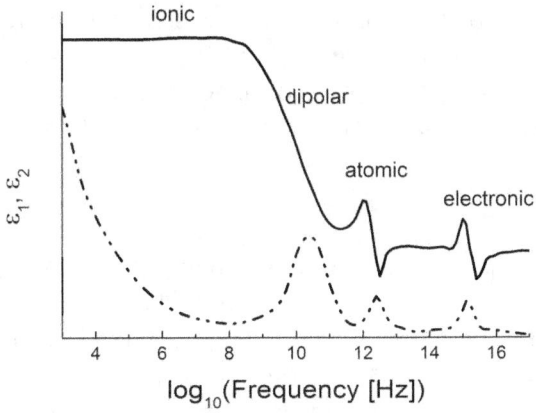

Fig. 3.2 Contributions of electronic, atomic, dipolar, and ionic excitations to the dielectric function

The harmonic oscillator model cannot always be applied. The biggest restriction results from the assumption of a linear response to the applied electric field. For large electric fields other interactions with matter can be expected and can be observed. Then, χ itself depends on **E**, i.e. $\chi(\mathbf{E})$. This field dependence can be described by a series expansion of the polarizability **P** in powers of **E**, where susceptibilities $\chi_{ijk\ldots}$ are introduced which now are a constant in relation to **E**:

$$P_i = \underbrace{\varepsilon_0 \sum_j \chi_{ij} E_j}_{\text{linear term } \chi^{(1)} \text{ (anisotropic medium)}} + \underbrace{\varepsilon_0 \sum_j \sum_k \chi_{ijk} E_j E_k}_{\text{quadratic term } \chi^{(2)}} + \underbrace{\varepsilon_0 \sum_j \sum_k \sum_l \chi_{ijkl} E_j E_k E_l}_{\text{cubic term } \chi^{(3)}} + \ldots$$

(3.9)

for each component i = x, y, z of **P**. For sufficiently small electric fields ($|\mathbf{E}| \leq 100$ V/m), which is fulfilled in most optical applications, the relation (3.9) is linear. In quantum mechanics this means that the electrons move in a parabolic band. Then, the susceptibility χ is a tensor of second rank, if the material is optically anisotropic. At large intensities the response of condensed matter on electric (and magnetic) fields becomes dependent on higher powers of the fields. This phenomenon is called *optical nonlinearity*, in contrast to the linear response at moderate intensities. In quantum mechanics optical nonlinearity is expressed by a movement in a nonparabolic band. Optical nonlinearities are not additive. However, they will not be subject of interest in the whole booklet.

3.1.2 The Drude Dielectric Function

In metals, semimetals, and semiconductors an important contribution to the dielectric function stems from unbound charge carriers or free carriers. In metals these are the *free electrons*. Within the harmonic oscillator model their contribution is obtained when assuming the resonance frequency $\omega_{fc} = 0$. In quantum mechanics this corresponds to a movement of the electrons in a potential $V(\mathbf{r}) = 0$. Then, the susceptibility of the unbound charge carriers - the *Drude susceptibility* [12, 13] - reads:

$$\chi_{fc}(\omega) = -\frac{\omega_p^2}{\omega^2 + i\omega\gamma_{fc}} \qquad (3.10)$$

with the abbrevation

$$\omega_p^2 = \frac{Ne_0^2}{Vm_{eff}\varepsilon_0} \qquad (3.11)$$

being the *plasma frequency of the unbound charge carriers* assuming them as a plasma. In a parabolic band structure the effective mass m_{eff} corresponds to the electron mass m_e, but in nonparabolic band structures m_{eff} may differ from m_e. For example, in silicon the negative free carriers in n-doped silicon have an effective mass of $m_{eff} = 0.8 \cdot m_e$ while the positive carriers in p-doped silicon have an effective mass of $m_{eff} = 0.26 \cdot m_e$.

Fig. 3.3 exemplarily shows the dielectric function and the corresponding refractive index of a free electron plasma with plasma frequency $5 \cdot 10^{15}$ s^{-1} and damping constant $5 \cdot 10^{14}$ s^{-1}.

For $\omega < \omega_P$ the real part ε_1 becomes negative and decreases rapidly with decreasing frequency. The imaginary part ε_2 is positive and increases with decreasing frequency. The refractive index n becomes small and increases only slowly with decreasing frequency while the absorption index κ rapidly increases.

The consequences of this behaviour are:
1.) Strong absorption of electromagnetic radiation in metals at $\omega < \omega_P$. The wave cannot penetrate far into the metal. Typical penetration

depths (skin depths) of real metals are in the order of 20 nm - 30 nm, depending on the metal.

2.) The reflection of a metal approaches almost 100%. For a negligible n and high κ-values the reflection at normal incidence approaches

$$R = \frac{(n-1)^2 + \kappa^2}{(n+1)^2 + \kappa^2} \approx \frac{1+\kappa^2}{1+\kappa^2} = 1. \qquad (3.12)$$

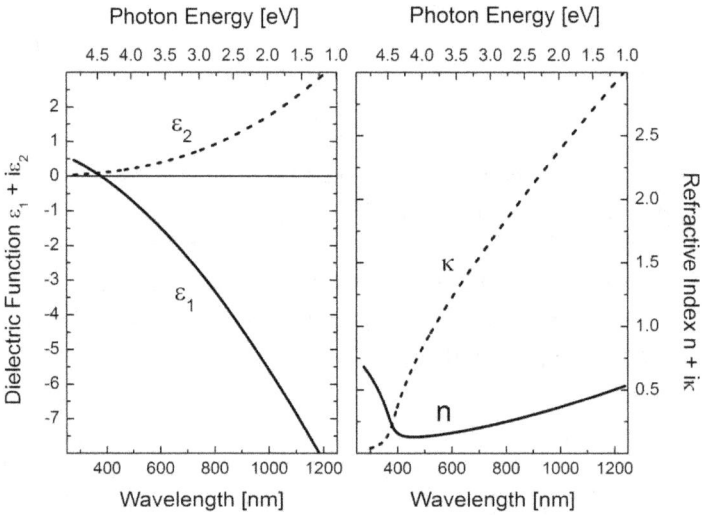

Fig. 3.3 Dielectric function and refractive index of a pure Drude metal with plasma frequency $5 \cdot 10^{15}$ s^{-1} and damping constant $5 \cdot 10^{14}$ s^{-1}

With the contribution of free charge carriers the dielectric function according to the harmonic oscillator model becomes

$$\varepsilon(\omega) = 1 - \underbrace{\frac{\omega_p^2}{\omega^2 + i\omega\gamma_{fc}}}_{\text{Drude susceptibility}} + \sum_n \underbrace{\frac{\omega_{Pn}^2}{\omega_n^2 - \omega^2 - i\omega\gamma_n}}_{\substack{\text{harmonic oscillators} \\ \text{(bound electrons)}}} \qquad (3.13)$$

At $\omega = 0$, $\varepsilon_1(0)$ still becomes constant but takes now large negative values depending on ω_P and γ_{fc} due to the contribution of the free charge carriers. The imaginary part $\varepsilon_2(\omega)$ always increases like $1/\omega$ with decreasing ω for small ω and diverges at $\omega = 0$. However, the conductivity $\sigma = \omega \cdot \varepsilon_2$ remains finite, approaching the DC-value

$$\sigma_{DC} = \varepsilon_0 \frac{\omega_p^2}{\gamma_{fc}}. \tag{3.14}$$

The damping constant γ_{fc} of the free charge carriers is closely related to the relaxation time τ, after which the common drift motion of the free charge carriers is relaxed by interactions with other charges and lattice defects:

$$\gamma_{fc} = \tau^{-1} = \tau^{-1}_{pointdefects} + \tau^{-1}_{dislocations} + \tau^{-1}_{grainboundaries} + \tau^{-1}_{surface/interface} + \tau^{-1}_{e-phonon} + \tau^{-1}_{e-e} \tag{3.15}$$

This *Mathiessen rule* only holds for independent relaxation processes, and hence only for a linear relation (3.1). Experimental values for τ are usually determined from the dc-conductivity (3.14).

3.1.3 Extensions of the Harmonic Oscillator Model

The harmonic oscillator model is quite often a good choice for the description of the dielectric function. However, sometimes mainly material properties give reason for an unsufficient description of the dielectric function with the harmonic oscillator model. For these cases several extensions were developed that take care of specific material properties. They are summarized below. Before we present a generalization of the harmonic oscillator stemming from Leng *et al.* [14]. The susceptibility of the *generalized harmonic oscillator* reads

$$\chi(\omega) = \frac{S}{\omega^2}\left[\frac{\exp(i\beta)}{(\omega_0-\omega-i\gamma)^m} + \frac{\exp(-i\beta)}{(\omega_0+\omega+i\gamma)^m}\right.$$
$$\left. -2\cdot\text{Re}\left(\frac{\exp(-i\beta)}{(\omega_0+i\gamma)^m}\right) + 2\cdot i\cdot m\cdot\omega\cdot\text{Im}\left(\frac{\exp(-i\beta)}{(\omega_0+i\gamma)^{m+1}}\right)\right]$$

(3.16)

where the oscillator has the resonance frequency ω_0 and a damping constant γ. The quantity β is an arbitrary phase and the power m is in general a positive real number. Re and Im mean the real part and the imaginary part of the corresponding expression.

Brendel and Kim oscillators

For *statistically pertubated* or *amorphous materials* extensions of the harmonic oscillator model exist as *Brendel oscillator* [15] or as *Kim oscillator* [16].

A *Brendel oscillator* is a harmonic oscillator with eigenfrequency ω_0 and width γ that is inhomogeneously broadened by an infinite sum over sharp harmonic oscillators with eigenfrequency x and width γ. These oscillators are Gaussian distributed around the harmonic oscillator with eigenfrequency ω_0 with a standard deviation σ:

$$\chi(\omega) = \frac{S\cdot\omega_0^2}{\sigma\cdot\sqrt{2\pi}}\int_0^\infty dx\, \frac{\exp\left(-0.5\cdot\left(\frac{x-\omega_0}{\sigma}\right)^2\right)}{x^2-\omega^2-i\omega\gamma}.$$

(3.17)

The advantage of a Brendel oscillator compared with a harmonic oscillator simply broadened by an increased γ is that the contures in the real and imaginary part of the dielectric function become smoother due to the Gaussian distribution.

In the simpler model of a *Kim oscillator* the damping γ of a harmonic oscillator with resonance frequency ω_0 is assumed to be frequency dependent:

$$\chi(\omega) = \frac{S \cdot \omega_0^2}{\omega_0^2 - \omega^2 - i\omega\gamma(\omega)} \tag{3.18}$$

with

$$\gamma(\omega) = \gamma \cdot \exp\left(-\frac{1}{1+\sigma^2}\left(\frac{\omega-\omega_0}{\gamma}\right)^2\right). \tag{3.19}$$

The parameter σ ≥ 0 is used to switch between a Gaussian or a Lorentzian shape of γ(ω). For σ = 0 a pure Gaussian shape is obtained, for σ > 5 a Lorentzian shape is obtained.

Tauc-Lorentz model, Cody-Lorentz model, and OJL model
Amorphous semiconductor and oxide materials often have optical functions that depend upon deposition conditions. Their optical constants also cannot be described by an unmodified harmonic oscillator. A first approach for the imaginary part of the dielectric function of these materials stems from Tauc et al. [17]. Jellison and Modine [18] derived a model based on a combination of the Tauc band egde and the Lorentz oscillator formulation, the *Tauc-Lorentz model*. An extension of the Tauc-Lorentz model is the *Cody-Lorentz model* [19]]. A further model for amorphous semiconductors is well known as *OJL model* from O'Leary, Johnson and Lim [20].
In the *Tauc-Lorentz model* the imaginary part of the complex dielectric function of amorphous materials with band gap (mainly semiconductor materials) can be modeled as

$$\varepsilon_{2,TL}(\omega) = \begin{cases} \dfrac{(\omega - \omega_{gap})^2}{\omega^2} \cdot \dfrac{S \cdot \omega_0^2 \cdot \gamma \cdot \omega}{(\omega^2 - \omega_0^2)^2 + \omega^2\gamma^2} & \omega > \omega_{gap} \\ 0 & \omega \leq \omega_{gap} \end{cases}$$

(3.20)

The oscillator has the resonance frequency ω_0 and a damping constant γ. ω_{gap} is the frequency corresponding to the band gap energy $E_{gap} = \hbar\omega_{gap}$. The real part $\varepsilon_{1,TL}$ is obtained from the imaginary part using Kramers-Kronig relations (see Sect. 3.1.4).

Note that in the original contribution of Jellison and Modine the authors used (photon) energies instead of circular frequencies. Then, (3.20) reads

$$\varepsilon_{2,TL}(E) = \begin{cases} \dfrac{A \cdot E_0 \cdot C \cdot (E - E_{gap})^2}{(E^2 - E_0^2)^2 + E^2 C^2} \cdot \dfrac{1}{E} & E > E_{gap} \\ 0 & E \leq E_{gap} \end{cases}$$

(3.21)

with $C = \hbar\gamma$. In comparison with (3.20) it follows that $A = S \cdot \hbar\omega_0$. That means that A need not become negative because neither the oscillator strength S nor the energy $\hbar\omega_0$ can become negative.

The dielectric function of a Tauc-Lorentz oscillator with resonance frequency $3.5 \cdot 10^{15}$ s^{-1}, damping constant $3.5 \cdot 10^{14}$ s^{-1}, oscillator strength S = 1, and band gap 1.5 eV is depicted in Fig. 3.4. In comparison to the harmonic oscillator (Lorentz oscillator) in Fig. 3.1 one can recognize that the imaginary part of the dielectric function as well as of the refractive index decreases more rapidly when going from the peak wavelength to longer wavelengths (lower photon energies). This is caused by the additional decrease according to the Tauc band edge. Moreover, for energies lower than the bandgap the imaginary parts are zero.

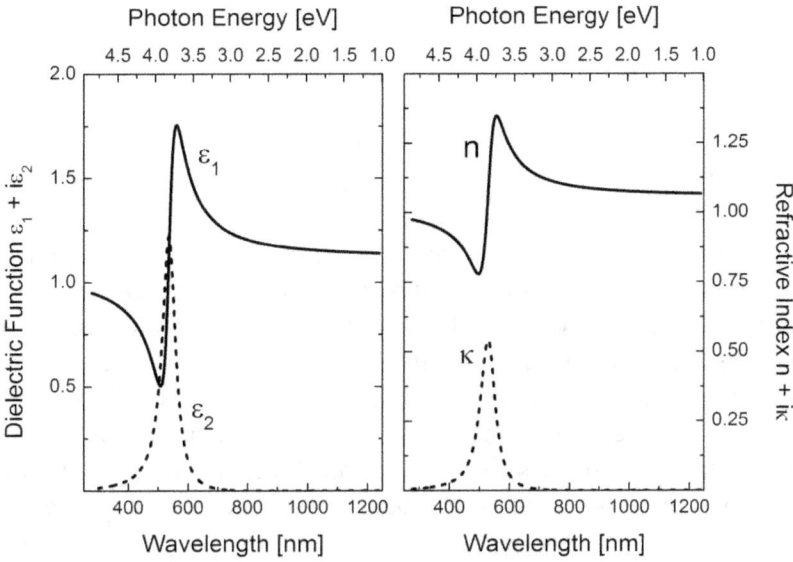

Fig. 3.4 Dielectric function of a Tauc-Lorentz oscillator with resonance frequency $3.5 \cdot 10^{15}$ s^{-1}, damping constant $3.5 \cdot 10^{14}$ s^{-1}, oscillator strength S = 1, and band gap 1.5 eV

Analogous to the Tauc Lorentz model Ferlauto *et al.* [19] developed the *Cody-Lorentz model*. The difference between this model and the Tauc Lorentz model is that up to a transition energy $E_1 = \hbar\omega_1$ the imaginary part of the dielectric function gets described by an Urbach-transition [21] with decay parameter Γ, and above this energy by a modified Tauc Lorentz approach. The modification is an additional transition at energy $E_2 + E_{gap}$ with $E_2 = \hbar\omega_2$ that separates the region where the absorption starts ($E < E_2 + E_{gap}$) from the harmonic oscillator behaviour ($E > E_2 + E_{gap}$). The value of T is automatically obtained from the continuity of both models at $E = E_1$. The real part $\varepsilon_{1,CL}$ is obtained from the imaginary part using Kramers-Kronig relations (see Sect. 3.1.4).

$$\varepsilon_{2,\text{CL}}(\omega) = \begin{cases} \dfrac{(\omega-\omega_{\text{gap}})^2}{(\omega-\omega_{\text{gap}})^2+\omega_2^2} \cdot \dfrac{S \cdot \omega_0^2 \cdot \gamma \cdot \omega}{(\omega^2-\omega_0^2)^2+\omega^2\gamma^2} & \omega > \omega_1 \\ \dfrac{T}{\omega}\exp\left(\dfrac{\omega-\omega_1}{\Gamma}\right) & 0 < \omega \leq \omega_1 \end{cases} \qquad (3.22)$$

Another model for the dielectric function of amorphous semiconductor materials stems from O' Leary, Johnson and Lim [20] and is well-known as *OJL model*. In a defect-free crystalline semiconductor, the absorption spectrum terminates abruptly at the energy gap. In contrast, in an amorphous semiconductor the absorption spectrum reaches into the gap region. The reason is that the electronic states in the valence band and conduction band can be divided into localized states and states which are randomly distributed through these amorphous semiconductors.

While the distribution of localized states follow a square-root functional dependence in the band region, the distribution shows an exponential functional dependence in the tail region with corresponding halfwidths γ_C and γ_V. This is illustrated in Fig. 3.5.

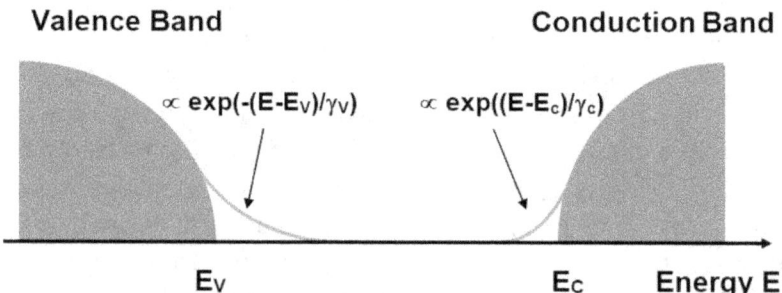

Fig. 3.5 Density of states N(E) in the model of O' Leary, Johnson and Lim [12] for an amorphous semiconductor

With this ansatz O' Leary, Johnson and Lim modeled the density of states for the conduction and valence band of an amorphous semiconductor. They derived the optical absorption coefficient $\alpha_{OJL}(E)$ with $E = \hbar\omega$ as

$$\alpha_{OJL}(E) = D^2(E) \frac{\sqrt{2}}{\pi^2 \hbar^3} m_C^{3/2} \frac{\sqrt{2}}{\pi^2 \hbar^3} m_V^{3/2} J(E) \quad (3.23)$$

The masses m_C and m_V are the effective masses of the electrons in the conduction band and the valence band. The density of states $D^2(E)$ is proportional to $\exp(-(E-E_{gap})/\Gamma)/E$ (Urbach transition) with a decay parameter Γ.

The normalized joint density-of-states (JDOS) $J(E)$ in the OJL-model differs for the two cases $E \leq E_{gap} + (\gamma_C+\gamma_V)/2$ and $E \geq E_{gap} + (\gamma_C+\gamma_V)/2$:

(1) $E \leq E_{gap} + (\gamma_C+\gamma_V)/2$

$$J(E) = \frac{\gamma_C^2}{\sqrt{2e}} \cdot \exp\left(\frac{E-E_{gap}}{\gamma_C}\right) \cdot Y\left(\frac{\gamma_V}{2\gamma_C}\right) + \frac{\gamma_V^2}{\sqrt{2e}} \cdot \exp\left(\frac{E-E_{gap}}{\gamma_V}\right) \cdot Y\left(\frac{\gamma_C}{2\gamma_V}\right)$$
$$+ \frac{1}{2\sqrt{e}} \frac{(\gamma_C\gamma_V)^{3/2}}{\gamma_V - \gamma_C} \cdot \left[\exp\left(\frac{E-E_{gap}-\gamma_C/2}{\gamma_V}\right) - \exp\left(\frac{E-E_{gap}-\gamma_V/2}{\gamma_C}\right)\right]$$
(3.24)

with $Y(z) = \sqrt{z} \cdot \exp(-z) + 0.5\sqrt{\pi} \cdot \text{erfc}(\sqrt{z})$ and erfc = error function.

(2) $E \geq E_{gap} + (\gamma_c + \gamma_v)/2$

$$J(E) = \frac{\gamma_c^2}{\sqrt{2}e} \cdot \exp\left(\frac{E - E_{gap}}{\gamma_c}\right) \cdot Y\left(\frac{E - E_{gap}}{\gamma_c} - \frac{1}{2}\right)$$
$$+ \frac{\gamma_v^2}{\sqrt{2}e} \cdot \exp\left(\frac{E - E_{gap}}{\gamma_v}\right) \cdot Y\left(\frac{E - E_{gap}}{\gamma_c} - \frac{1}{2}\right) + \quad (3.25)$$
$$+ \frac{1}{2\sqrt{e}}(E - E_{gap})^2 \cdot L\left(\frac{\gamma_c}{2(E - E_{gap})}, \frac{\gamma_v}{2(E - E_{gap})}\right)$$

with $L(x, y) = \int_x^{1-y} \sqrt{z}\sqrt{1-z}\, dz$.

The imaginary part of the complex dielectric function of amorphous semiconductor materials with band gap E_{gap} finally follows from the optical absorption coefficient $\alpha_{OJL}(E)$ as

$$\varepsilon_{2,OJL}(E) = \frac{\hbar c \cdot \alpha_{OJL}(E)}{E}. \quad (3.26)$$

The real part $\varepsilon_{1,OJL}$ is obtained from the imaginary part by Kramers Kronig relations.

Extended Drude susceptibility

The classical Drude model (see Sect. 3.1.2) works with a damping constant that does not depend upon frequency. This is a good approximation in most cases. However, there are situations where the damping of the free carriers exhibits a characteristic dependence on frequency, for example in the case of scattering at charged impurities. The *extended Drude model* [22] uses a rather simple, but successful choice of the frequency dependence of the damping term:

$$\gamma_{fc}(\omega) = \Gamma_L - \frac{\Gamma_L - \Gamma_H}{\pi}\left[\tan^{-1}\left(\frac{\omega - \Omega_\Gamma}{\Gamma_{\Gamma\omega}}\right) + \frac{\pi}{2}\right]. \quad (3.27)$$

The function for the damping constant is chosen to change smoothly from a constant Γ_L at low frequencies to another constant level Γ_H at high frequency. The transition region is defined by the crossover frequency Ω_Γ -the center of the transition region- and the crossover width parameter $\Gamma_{\Gamma\omega}$.

Forouhi and Bloomer

A controversial model is the model of Forouhi and Bloomer [23] for amorphous materials. It gives a relation for the absorption index κ in dependence on the photon energy E rather than the imaginary part ε_2 of the dielectric function

$$\kappa_{FB}(E) = \sum_{j=1}^{N} \frac{A_j(E - E_{gap})^2}{E^2 - B_j E + C_j}, \quad (3.28)$$

with A_j, B_j, and C_j being positive, nonzero constants characteristic of the medium, so that $4C_j - B_j^2 > 0$. E_{gap} is the band gap energy. The corresponding refractive index n_{FB} is obtained by a Hilbert transform

$$n_{FB}(E) = n_{FB}(\infty) + \frac{1}{\pi}\wp\int_{-\infty}^{\infty}\frac{k_{FB}(\Omega) - k_{FB}(E)}{\Omega - E}\cdot d\Omega, \quad (3.29)$$

where $\wp$ is the principal value of the integral, to

$$n_{FB}(E) = n_{FB}(\infty) + \sum_{j=1}^{N}\frac{B_{0j}E + C_{0j}}{E^2 - B_j E + C_j}. \quad (3.30)$$

B_{0j} and C_{0j} are constants that depend on A_j, B_j, C_j, and $n_{FB}(\infty)$ is a constant greater than unity.

There are some problems with these formulas:

- For certain parameters A, B, C the absorption index may be less than zero, $k_{FB}(E) < 0$, for $E < E_{gap}$. This is unphysical.
- $k_{FB}(E)$ becomes constant for $E \to \infty$. In experiments, $k(E) \to 0$ for $E \to \infty$ proportional to $1/E^3$.
- The time-reversal symmetry requires that $k(-E) = -k(E)$, which is not satisfied.
- As $k_{FB}(\infty) \neq 0$, Kramers Kronig relations are not fulfilled.

3.1.4 Kramers Kronig Relations

The real and imaginary part of the dielectric function are related in a way so that if one knows one of them the other can calculated from it. The relations are the *Kramers-Kronig relations* [24, 25] sometimes called *dispersion integrals*.

Assuming that $\chi(\omega)$ vanishes at very high frequencies, i.e $\lim_{\omega \to \infty} \chi(\omega) = 0$, one finds

$$\varepsilon_1(\omega) = 1 + \frac{2}{\pi} \wp \int_0^\infty \frac{\Omega \cdot \varepsilon_2(\Omega)}{\Omega^2 - \omega^2} d\Omega \qquad (3.31)$$

$$\varepsilon_2(\omega) = -\frac{2}{\pi} \wp \int_0^\infty \frac{\varepsilon_1(\Omega) - 1}{\Omega^2 - \omega^2} d\Omega \qquad (3.32)$$

where $\wp$ is the principal value of the integral.

These relations are not restricted on (ε_1, ε_2) but hold for any frequency-dependent function that connects an output with an input in a linear causal way. Therefore, they are often used to replace the measurement of one part of a complex function like the dielectric function by a corresponding calculation of the other part. For example, many published optical constants result from a Kramers-Kronig analysis of the reflectance or the absorption coefficient or from combination of electron energy-loss experiments with Kramers-Kronig relations.

Problems with the Kramers-Kronig relations usually arise from the fact that the integrals for the real part and the imaginary part are extended from 0 to ∞. Experimental values are, however, only available for a restricted interval [Ω_1, Ω_2]. Therefore, the above integrals must be divided into parts, for example the real part of the dielectric function ε_1 can be written as

$$\varepsilon_1(\omega) = 1 + \frac{2}{\pi}\int_0^{\Omega_1}\frac{\Omega\cdot\varepsilon_2(\Omega)}{\Omega^2-\omega^2}d\Omega + \frac{2}{\pi}\wp\int_{\Omega_1}^{\Omega_2}\frac{\Omega\cdot\varepsilon_2(\Omega)}{\Omega^2-\omega^2}d\Omega + \frac{2}{\pi}\int_{\Omega_2}^{\infty}\frac{\Omega\cdot\varepsilon_2(\Omega)}{\Omega^2-\omega^2}d\Omega$$

(3.33)

While the mid integral in the interval [Ω_1, Ω_2] can be calculated exactly, the two other integrals must be estimated using reasonable assumptions. For this purpose various authors supposed different extrapolation functions. The most common and the oldest of these is Roessler's extrapolation [26, 27]. Roessler applied the Kramers-Kronig relations on reflectance measurements and evaluated the contributions from the outside regions by means of the mean value theorem asserting that there exists a frequency Ω_{low} in the interval [0, Ω_1] and one Ω_{high} in the interval [Ω_2, ∞] at which the integrals yield the required contribution to the total integral.

Riu and Lapaz [28] discussed limitations and errors of the Kramers-Kronig relations. They concluded that the Kramers-Kronig relations were practically applicable in almost every experimental situation.

3.2 Empirical Models for the Refractive Index

Optical constants n+iκ or $\varepsilon_1+i\varepsilon_2$ can always be modeled using the physical models presented in Sect. 3.1. However, especially glass manufacturers developed and still use today empiric formulas for the refractive index n(λ) to parametrize the precise measurements at distinct well-defined wavelengths. In this Sect. we give an overview on the mostly used empiric formulas for the refractive index.

One of the most prominent formula is the *Sellmeier formula* [29].

$$n^2 - 1 = \sum_{j=1}^{N} \frac{A_j \lambda^2}{\lambda^2 - B_j} . \qquad (3.35)$$

This formula is the most physical since it corresponds to the sum over j = 1, ..., N undamped harmonic oscillators with eigenfrequencies $\omega_j = \frac{2\pi c}{\sqrt{B_j}}$ and oscillator strengths $S_j = A_j$.

This formula is nowadays used from glass manufacturers like SCHOTT AG (since 1992), and OHARA Inc. as three-term Sellmeier formula to approximate the refractive index of their glass. The most important modification of this formula is to replace n^2-1 by $n^2 - n_0^2$.

Another often used formula is the *Schott formula*. Originally developed by Erich Schott at SCHOTT AG in 1966 and also used by SCHOTT AG until 1992, it is nowadays used from other glass manufacturers like CORNING Inc., HOYA Inc., HIKARI Inc., or SUMITA Inc. The general form is

$$n^2 = \sum_{j=0}^{N} A_j \lambda^{2j} + \sum_{k=1}^{M} B_k \lambda^{-2k} . \qquad (3.36)$$

The original Schott formula is obtained for N = 1 and M = 4. The Schott formula is also known as *Laurent formula*, because it corresponds to a Laurent series in the wavelength λ.

The third important and often used empiric formula is the *Cauchy formula* from the prominent mathematician A. L. Cauchy [30, 31]:

$$n = A_n + \frac{B_n}{\lambda^2} + \frac{C_n}{\lambda^4} \qquad \kappa = A_\kappa + \frac{B_\kappa}{\lambda^2} + \frac{C_\kappa}{\lambda^4}. \qquad (3.37)$$

The advantage of the Cauchy formula compared to Sellmeier and Schott formula is that it also considers the imaginary part κ of the complex refractive index. It is therefore suited to fit also the optical constants of absorbing materials. It is often used for photoresists which are absorbing in the UV and at wavelengths in the violet/blue visible spectral region.

In close relation to the Cauchy formula another formula was developed, the *exponential Cauchy formula*:

$$n = A_n + \frac{B_n}{\lambda^2} + \frac{C_n}{\lambda^4} \qquad \kappa = A_\kappa \exp\left(B_\kappa \left(\frac{1.239841875}{\lambda}\right) - C_\kappa\right). \qquad (3.38)$$

The difference to the Cauchy formula is in the ansatz for the absorption index κ, which now is described by an exponential function. The value 1.239841875 is valid for wavelengths in µm.

In fact, many optical constants can be approximated by these few empiric formulas. Besides these formulas, a series of more or less known empiric formulas exist that are applied to transparent materials. They are comprised in the following Table 3.1 together with the Sellmeier, Schott, Cauchy, and exponential Cauchy formula.

Table 3.1 Empirical formulas for the complex refractive index. Wavelength in μm.

Name	Formula
Sellmeier [29]	$n^2 - 1 = \sum_{j=1}^{N} \dfrac{A_j \lambda^2}{\lambda^2 - B_j}$ $n^2 - n_0^2 = \sum_{j=1}^{N} \dfrac{A_j \lambda^2}{\lambda^2 - B_j}$ Original Sellmeier formula: N = 3. Mostly used with N = 2.
Schott	$n^2 = \sum_{j=0}^{N} A_j \lambda^{2j} + \sum_{k=1}^{M} B_k \lambda^{-2k}$ Original Schott formula for N = 1 and M = 4. Formulas with N ≤ 4 and M ≤ 5 are used.
Cauchy [30, 31]	$n = A_n + \dfrac{B_n}{\lambda^2} + \dfrac{C_n}{\lambda^4},$ $\kappa = A_\kappa + \dfrac{B_\kappa}{\lambda^2} + \dfrac{C_\kappa}{\lambda^4}$
exponential Cauchy	$n = A_n + \dfrac{B_n}{\lambda^2} + \dfrac{C_n}{\lambda^4}$ $\kappa = A_\kappa \exp\left(B_\kappa \left(\dfrac{1.239841875}{\lambda} \right) - C_\kappa \right)$
Conrady [32, 33]	$n = A + \dfrac{B}{\lambda} + \dfrac{C}{\lambda^{3.5}},$ $n = A + \dfrac{B}{\lambda^2} + \dfrac{C}{\lambda^{3.5}}$ The first formula is used more often.
Kingslake [34]	$n^2 = A + \sum_{j=1}^{N} \dfrac{B_j}{\lambda^2 - C_j}$ This formula is known as Kettler-Drude or Helmholtz-Drude formula for N = 2.

Herzberger [35, 36]	$$n = A + \sum_{j=1}^{N} B_j \lambda^{2j} + \sum_{k=1}^{M} \frac{C_k}{(\lambda^2 - 0.028)^k}$$ The Herzberger formula is used at wavelengths in the infrared. Mostly used are N = 2 or 3 and M = 2. A further modification is to exchange 0.028 with 0.035 or any other value.
Hartmann [37]	$$n = A + \frac{B}{(\lambda - C)^N}$$ There exist three variations: N = 1, N = 2, and N = 1.2.
modified Sellmeier [2]	$$n^2 = A + \sum_{j=1}^{N} B_j \lambda^{2j} + \sum_{k=1}^{M} C_k \lambda^{-2k} +$$ $$+ \sum_{p=1}^{P} \frac{D_p \lambda^2}{\lambda^2 - E_p} + \sum_{q=1}^{Q} \frac{F_p}{\lambda^2 - G_p}$$ $$n^2 - n_0^2 = \frac{A_1 \lambda^2}{\lambda^2 - B_1} + \frac{A_2}{\lambda^2 - B_2}$$ Different combinations of Sellmeier, Schott and Herzberger formula are in use, depending upon N, M, P, and Q. Special case: $A = n_0^2$, N = M = 0, P = Q = 1
Handbook of Optics [2]	$$n^2 = A + \frac{B}{\lambda^2 - C} - D\lambda^2$$ $$n^2 = A + \frac{B\lambda^2}{\lambda^2 - C} - D\lambda^2$$ Were introduced in the Handbook of Optics. Can be derived from the modified Sellmeier formula for N = 1, M = P = 0, Q = 1, and for N = 1, M = Q = 0, and P = 1.

3.3 Effective Medium Approaches

This Sect. is concerned with the optical constants of a special class of materials: composites of at least two nonmixable components. Usually, these are well-separated inclusions statistically distributed in a nonabsorbing homogeneous matrix (see Fig. 3.6). The optical response of such a medium is determined by the inclusions as well as the matrix material and it is difficult to predict it in general. However, if it is possible to replace the inhomogeneous composite by a homogeneous material of one common dielectric function ε_{eff}, the reflectance, transmittance, and absorbance of this medium can be calculated as linear response. For that purpose, a model for the dielectric function ε_{eff} of this *effective medium* must be established in dependence on the inclusion properties and the surrounding matrix and the concentration of inclusions in the composite.

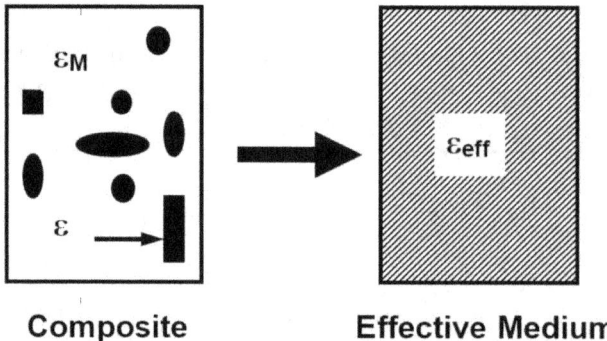

Composite **Effective Medium**

Fig. 3.6 Scheme of the effective medium: The realistic composite is replaced by a homogeneous effective medium

The fundamental question is how to get this effective dielectric function ε_{eff} in terms of the dielectric function ε of the inclusions and ε_M of the matrix and of suitably chosen topology parameters. A considerable number of formulas for the effective dielectric function is available today based upon different approximative models. The most important will be reviewed in this chapter.

A fundamental assumption which is also the most limiting assumption of all *effective medium theories* is that scattering by the inclusions can be neglected. This is a strong condition, since always scattering occurs even for nanometer sized inclusions. Additionally, scattering depends upon the material properties of the inclusions, not only on size and shape. Hence, the application of EMA models on composites with transparent purely scattering inclusions is questionable as well as for composites with strongly absorbing inclusions, since it can be proven that the scattering by absorbing inclusions of same size and shape as nonabsorbing inclusions is larger by a factor of 10 - 100.

A further restricting assumption is that the inclusions must be well-separated. Electromagnetic particle-particle interactions usually are omitted. This limits the concentration of the inclusions on less than 2% of the total volume, as was shown in [9, 38].

The first effective medium concept goes back to Newton (see [39]). It has been modified by Beer [40], Gladstone and Dale [41], Landau and Lifschitz [42], and Looyenga [43]. It is based on simply averaging certain powers of the dielectric functions of the two mixed components, weighted with the filling factor f:

$$\varepsilon_{eff} = f\varepsilon + (1-f)\varepsilon_M \qquad \text{Newton} \qquad (3.39)$$

$$\varepsilon_{eff}^{1/2} = f\varepsilon^{1/2} + (1-f)\varepsilon_M^{1/2} \qquad \text{Beer, Gladstone} \qquad (3.40)$$

$$\varepsilon_{eff}^{1/3} = f\varepsilon^{1/3} + (1-f)\varepsilon_M^{1/3} \qquad \text{Landau-Lifschitz, Looyenga} \qquad (3.41)$$

The *filling factor* f or *volume fraction* is defined as the ratio of the mass concentration of inclusions m/V and the mass density ρ of the inclusions:

$$f = \frac{m}{V} \cdot \frac{1}{\rho}. \tag{3.42}$$

Lichtenecker [39] obtained a *logarithmic mixture rule*

$$\log(\varepsilon_{eff}) = f \cdot \log(\varepsilon) + (1-f) \cdot \log(\varepsilon_M) \tag{3.43}$$

The simplest approach to an effective medium that explicitely considers also the shape of the inclusions stems from J.C. Maxwell Garnett [44] in 1904 for spherical inclusions

$$\varepsilon_{eff} = \varepsilon_M \frac{(\varepsilon + 2\varepsilon_M) + 2f(\varepsilon - \varepsilon_M)}{(\varepsilon + 2\varepsilon_M) - f(\varepsilon - \varepsilon_M)} \quad \text{or} \quad \frac{\varepsilon_{eff} - \varepsilon_M}{\varepsilon_{eff} + 2\varepsilon_M} = f \frac{\varepsilon - \varepsilon_M}{\varepsilon + 2\varepsilon_M} \tag{3.44}$$

In the Bruggeman-ansatz [45] the dielectric constant is given from

$$f \cdot \frac{\varepsilon - \varepsilon_{eff}}{\varepsilon + 2\varepsilon_{eff}} + (1-f) \cdot \frac{\varepsilon_M - \varepsilon_{eff}}{\varepsilon_M + 2\varepsilon_{eff}} = 0 \tag{3.45}$$

or, resolved for ε_{eff} from

$$\varepsilon_{eff} = E \pm \sqrt{E^2 + \frac{\varepsilon \varepsilon_M}{2}} \tag{3.46}$$

with

$$E = \frac{\varepsilon_M \cdot (2-3f) - \varepsilon \cdot (1-3f)}{4}. \tag{3.47}$$

Usually, the positive square root is used. However, if the imaginary part of ε_{eff} becomes negative, the negative square root has to be used to obtain the correct dielectric function. For more than two components of the composite the Bruggeman solution is

$$\sum_n f_n \cdot \frac{\varepsilon_n - \varepsilon_{eff}}{\varepsilon_n + 2\varepsilon_{eff}} = 0. \tag{3.48}$$

All these results for ε_{eff} can be used to calculate the reflectance, transmittance and absorbance of the composite using Maxwell's relation between the dielectric function and the refractive index, Fresnel's equations, and the corresponding formula for the absorption coefficient.

Extensions of existing effective medium theories were often made by several authors. The most important are the extensions on two dimensions [46]. For this replace the 2 in the denominators of Eqs. (3.43), (3.44), and (3.47) by (D-1). D = 3 yields the three-dimensional results, for D = 2 the two-dimensional results are obtained.

Although the restrictions of effective medium models are considerable, EMAs have some applications. One is the calculation of the refractive index of dielectric particle – dielectric matrix composites. Inclusion of high refractive nanoparticles results in an increased refractive index. Vice versa, the inclusion of low refractive-nanoparticles with a refractive index lower than that of the matrix material lowers the refractive index of the composite. Nevertheless, we want to point to the fact that the mean haze becomes intolerable according to the ASTM standard D 1003 - 97 *Test Method for Haze and Luminous Transmittance of Transparent Plastics* for filling factors $f \geq 0.01$ and high refractive nanoparticles. Another application is the consideration of roughness in ellipsometry of thin film systems. The rough surface is treated as a homogeneous thin layer with effective refractive index. For the calculation of the effective refractive index EMA models are used with the dielectric functions of the surface material and the material in front of the surface.

4 The Practical Determination of Optical Constants

For the determination of optical constants we can distinguish between *direct methods* and *indirect methods*. The most methods are indirect that means they measure one or two quantities that depend in a mostly nonlinear way upon the optical constants and then derive the optical constants using calculations, models and fits.

We further can distinguish between experiments where the optical constants of a thick substrate (transparent or opaque) are determined or where the optical constants of a layer on a substrate or even a layer stack shall be determined. The latter becomes increasingly important as in many applications with new materials, e.g. organic solar cells, organic LEDs, fuel cells, photolithography, and some more, a simple and fast determination of optical constants is desired and the material is coated on a substrate.

In the following we give an overview on some commonly used direct and indirect methods. This overview may be incomplete because we want to focus on indirect spectral measurement methods and the determination of optical constants from these spectral measurements.

The number of *direct methods* for determination of optical constants is pretty small and mostly allows the determination of the refractive index.

Deviation Angle Method

With this method the refractive index n of a highly transparent solid material for which the absorption index κ is negligible small can be determined. It is well described in several textbooks on optics [47-50]. It relies on the accurate measurement of the minimum deviation angle θ_{min} of an isosceles triangular prism made of the transparent material placed in air. From Snell's law the refractive index can be estimated by

$$n = \frac{\sin\left(\frac{\theta_{min} + \delta}{2}\right)}{\sin\left(\frac{\delta}{2}\right)} \cdot n_{air} \qquad (4.1)$$

where δ is the refracting angle (apex angle) of the prism and n_{air} the refractive index of air.

Often glass manufacturers use this method to determine very accurately the refractive index of their produced glasses at certain well-known wavelengths. Then, these data are used in a fit with one of the empiric formulas Sellmeier or Schott to find according coefficients. With the coefficients the dispersion of the glass can be calculated over a broad wavelength range.

Prism Coupler
Prism coupling [51-54] is a method that utilizes the total internal reflection of a light beam (mostly a laser beam) at the base of a prism to generate an evanescent wave that couples under certain conditions into the medium put in front of the prism. If the evanescent wave couples into this medium the internal total reflection gets attenuated what can be detected with a photodetector that measures the intensity of the totally reflected light.

For understanding of this method one must consider the thin film as planar waveguide in which nonradiating eigenmodes can propagate along with the film. This presumes that the film has a higher refractive index than the bordering media. These propagating nonradiating eigenmodes can be described by waves that propagate in the film with effective refractive index n_{eff} but decrease exponentially in their amplitude in the bordering.

The solutions differ for TE-polarization and TM-polarization. If the propagation constant of the evanescent wave coincides with the propagation constant of any of the waveguide modes the evanescent waves couples into the waveguide, leading to a drastic reduction of the totally reflected intensity at the prism base. At least, there must be two angles θ where the total reflection gets attenuated to solve for the film refractive index and the thickness simultaneously. One angle allows only for the determination of the refractive index. If three or more mode lines are observable, also a mean and standard deviation calculation is possible.

The angle θ is a function of the angle of incidence α, the apex angle δ of the prism, and the refractive index of the prism n_P

$$\theta = \delta + \sin^{-1}\left(\frac{\sin\alpha}{n_P}\right) \tag{4.2}$$

so that it is sufficient to record the intensity of the totally reflected light versus the incident angle.

Precise accuracy and resolution values depend on film type, thickness range, and rotary table resolution. Typical values for commercial prism couplers are ±(0.5 % + 5 nm) for the thickness accuracy and ±0.3 % for the thickness resolution. For the additional refractive index determination the accuracy is ±0.001 with a resolution of ±0.0005. With standard prisms, films and bulk materials with refractive index 3.35 and below are measurable. The thickness range is from 0.2 μm to 150 μm.

Absorbance measurement

Absorbance measurement can be used to determine the absorption index or extinction coefficient κ. Passing through an absorbing medium of thickness d, we can observe that the intensity I_0 gets exponentially attenuated

$$I(d) = I_0 \cdot \exp(-\alpha \cdot d) \tag{4.3}$$

according to *Lambert-Beer's law*. The factor α in the exponential function is the *absorption coefficient*

$$\alpha(\lambda) = \frac{4\pi}{\lambda} \kappa(\lambda). \tag{4.4}$$

Measuring two transmittances $T(d_1)$ and $T(d_2)$ with $d_2 > d_1$ at the same wavelength, the extinction coefficient follows as

$$\kappa(\lambda) = \frac{\lambda}{4\pi} \frac{\log(T(d_1)) - \log(T(d_2))}{((d_2 - d_1))} \qquad (4.5)$$

Other Direct Methods
Direct determination of the refractive index n can be carried out also with either one reflectance measurement at two angles or one reflectance measurement at one angle and one reflectance measurement at a special angle. There are two special angles that are used:
(a) the principal angle of incidence,
(b) the pseudo Brewster angle.
The idea of the principal angle is to illuminate the sample with linearly polarized under an angle of incidence of 45°. The reflected light is usually elliptically polarized with an arbitrary phase difference. Now, the angle of incidence gets changed until the phase difference between the p- and s-polarized component is 90°. The angle of incidence where this takes place is called principal angle of incidence.
The pseudo Brewster angle is the angle of incidence where the p-polarized reflected light vanishes (Brewster angle) or has a minimum (pseudo Brewster angle).

The most methods for determination of optical constants are *indirect*. They measure one or two quantities that depend in a nonlinear way upon the optical constants and from which the optical constants are derived using calculations, models and fits. There are mainly the following classes of experiments

- measuring spectral reflectance, transmittance, or absorbance,
- ellipsometry,

- Fourier transform spectroscopy methods,
- Mach-Zehnder interferometry,
- Attenuated Total Reflection (ATR),
- photothermal / photoacoustic spectroscopic methods (absorbance), and
- Angle-of-Incidence Total-External-Reflectance Method.

Fourier transform spectroscopy methods in principle also measure reflectance, transmittance, or absorbance using interferometric methods in the mid and far infrared spectral range. Recent developments showed that it also can be extended to the visible and ultraviolet spectral range. These methods determine phase and amplitude at many wavelengths simultaneously. Measurements are usually made at normal incidence with the specimen in one arm of a Michelson interferometer.

While in a Michelson interferometer the interference is established by shifting of a mirror, the interference conditions in a *Mach-Zehnder interferometer* are changed when a sample is set into one arm of the interferometer.

The *Attenuated Total Reflection* (ATR) method operates by measuring the changes in the total internal reflection of an infrared beam by coupling of the evanescent wave generated in an optically dense crystal in direct contact with the sample. In regions where the sample material absorbs, the total reflection gets altered by the attenuation of the evanescent wave.

Photothermal / photoacoustic spectroscopic methods are used to measure optical absorption in homogeneous media. Rather than transmission spectroscopic techniques, these techniques measure the absorption directly, thereby achieving high accuracy.

Total external reflection generally occurs when light is incident onto a material with the real part n of the refractive index less than unity.

This happens at high photon energies in the XUV and X-ray spectral region. Similar to the total internal reflection also here a critical angle of incidence θ_c exists where total external reflection starts. Then, the complex refraction index can be quite accurately by carrying out measurements of reflection versus angle of incidence. This is the *Angle-of-Incidence Total-External-Reflectance Method*.

A special technique was developed by Swanepoel [55]. His *envelope method* is based upon the transmittance measurement of a thin transparent film on a substrate. If the film thickness of the semi-transparent film is thick enough so that layer thickness interferences can be observed in the transmittance spectrum, the optical constants can be retrieved from various envelopes in the spectrum.

In the following we will concentrate on two easily accomplishable methods: spectral reflectance measurement and spectral ellipsometry.

4.1 Spectral Reflectance Measurement

Reflectance measurements are power measurement techniques. That means that the intensity reflectivity coefficients R_p and R_s are measured rather than the complex magnitude reflectivity coefficients r_p and r_s (Fresnel coefficients). The relation between them is

$$\left|r_{p,s}(\lambda)\right|^2 = R_{p,s}(\lambda). \tag{4.6}$$

Reflectance measurements can be conducted under normal incidence (angle of incidence $\alpha = 0°$) or under oblique incidence.
For *normal incidence* one can ignore polarization effects since most materials are rotationally symmetric. Thus, the spectral reflectance measurement can be realized without moving parts, resulting in simpler low cost instruments. For normal incidence the complex Fresnel coefficient $r(\lambda)$ of the surface of a bulk material in ambient air or better vacuum is directly connected with the complex refractive index $N(\lambda) = n(\lambda) + i \cdot \kappa(\lambda)$

$$r(\lambda) = |r(\lambda)| \cdot \exp(i \cdot \theta(\lambda)) = \frac{N(\lambda) - 1}{N(\lambda) + 1}. \qquad (4.7)$$

However, the determination of $N(\lambda)$ is rendered more difficult since the phase $\theta(\lambda)$ must be measured besides the reflectance $R(\lambda)$. For wavelengths less than 200 nm phase discriminating components are strongly absorbing. Then, the measurement of the phase becomes almost impossible. Instead, the measurement of the reflectance is supplemented by a mathematical method, the already introduced Kramers-Kronig relations (see Sect. 4.1.4). Applying the natural logarithm on Eq. (4.7) we get

$$\log(r(\lambda)) = \frac{1}{2}\log(R(\lambda)) + i \cdot \theta(\lambda). \qquad (4.8)$$

It can be proven that $\log(R)$ and θ satisfy the conditions of the Kramers Kronig relations. Then, θ can be calculated from $\log(R)$ and both can be used to determine n and κ. For solid materials the relations for n and κ are

$$n(\lambda) = \frac{1 - R(\lambda)}{1 + R(\lambda) - 2\sqrt{R(\lambda)} \cdot \cos(\theta(\lambda))}, \qquad (4.9)$$

$$\kappa(\lambda) = \frac{-2\sqrt{2R(\lambda)} \cdot \sin(\theta(\lambda))}{1 + R(\lambda) - 2\sqrt{R(\lambda)} \cdot \cos(\theta(\lambda))}. \qquad (4.10)$$

The most important problem encountered in the application of Kramers Kronig relations is to calculate the contributions arising from outside the measuring range (see again 3.1.4 Kramers Kronig relations).

For *oblique incidence* several methods are available [56]:

a) R_s and R_p at one angle of incidence,
b) reflectance measurements at two angles of incidence using natural or polarized radiation,
c) the ratio R_p/R_s at two angles of incidence,
d) pseudo Brewster angle and R_s or R_p at that angle,
e) pseudo Brewster angle and R_p/R_s at that angle,
f) pseudo Brewster angle and R_s, R_p, or R_p/R_s at any other angle of incidence,

All these methods require a rather accurately trimmed mechanical setup to achieve a high accuracy in defining the angle of incidence respectively the angle of reflection and in determination of the pseudo Brewster angle. Therefore, these methods yield results of variable accuracy. Methods a.) and b.) suffer markedly from a lack of sensitivity, while method c.) seems to be more suitable since the ratio R_p/R_s is high sensitive on changes of the optical constants of the solid.

A special setup for absolute measurement of R_s and R_p at one angle of incidence is the *VW optics* or *VW method* (see Fig. 4.1) according to Strong [57]. Modified versions have been proposed by several authors [58-61].

In the so-called V-mode the instrument beam is interacting with three mirrors (M1, M2 and M3). In the W-mode the beam additionally interacts twice with the sample. The ratio of the two readings produces the square of the sample reflectance.

The common way is to use the instrument under near-normal incidence, i.e. with an angle of incidence of 8 degrees. But it can also be used under oblique incidence. The near-normal incidence has the advantage that for almost all materials the reflectivities in p- and s-polarization are practically identical, $R_p = R_s$.

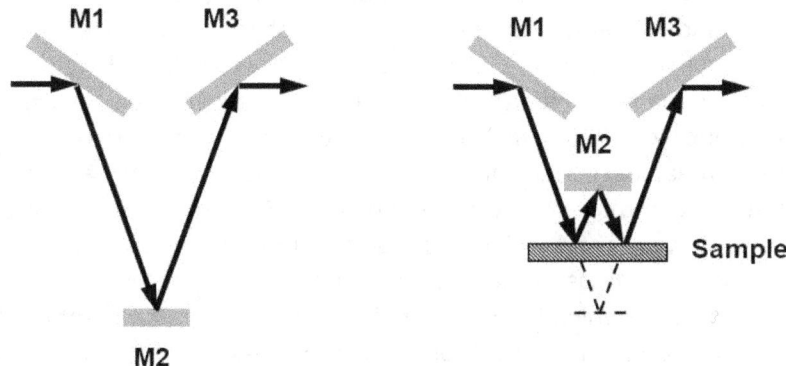

Fig. 4.1 Schematic configurations of the VW optics. At left the reference measurement with three mirrors M1, M2, and M3. At right the sample measurement

Preparing a material for reflectance measurements from which n and κ are to be determined often requires vacuum conditions. This is right particularly for metals and semiconductors. The reason is the chemical reactivity of most of the metals and semiconductors. Chemical reaction with oxygen leads to very thin passivating oxide layers that disturbs the reflectance measurement. This oxidation and further other reactions with other gaseous components must be avoided. This is one reason for many experimental data on metals are obtained from measurements at synchrotron radiation facilities, e.g. in the comprehensive database [62].

On the other hand, in many modern applications the material for which n and κ are to be determined is coated as a thin film on a substrate. This film has been prepared using several preparation techniques (CVD, PVD, sol-gel method, sputtering, evaporation, etc.) and its optical constants shall be determined preferably *in-situ*, in a short time, and under ambient conditions. The problem arising with the thin film is that it almost always exhibits layer thickness oscillations in the reflectance that superpose the properties of the bulk surface. Then, the application of the above methods becomes difficult, since no simple or clear expression can be written for these cases. The solution for this problem consists of a simple and rugged measurement setup that can be adjusted to the specific environmental conditions and a highly sophisticated mathematical processing of the data, using models for the layer stack as well as for the optical constants. In the following we will concentrate on reflectometric measurements with miniaturized spectrometers and their evaluation with respect to optical constants. Spectroscopic ellipsometric measurements and their evaluation are subject of the next Sect. 4.2.

Quick inline measurement during a production process requires a measuring setup which is robust and is adjusted to the ambient conditions at the production site. Moreover, the measured results must be stable and reproducible for a long time. These requirements can be easily fulfilled with a setup for spectral reflectance measurement that utilizes high-valued miniaturized spectrometer modules. They are extremely robust and have an excellent long-term stability of the wavelength calibration. Additionally, one can carry out multiplexed measurements at several sites with only one spectrometer system, for example when using fiber multiplexers.

For a reflectometric measurement a light source, an optical fiber, and optionally a measuring head are used to illuminate the sample with unpolarized light. The reflected light gets collected by the measuring head (optionally) and a second fiber which is connected to the detector. The direction of incidence may include an angle α with respect to the normal on the sample, but usually this angle is $\alpha = 0°$ (normal incidence). The principal setup for a reflectometric measurement is sketched in Fig. 4.2.

Typically a so-called Y-fiber is used, where two separate fibers are assembled so that the branch for the illumination and the branch for the detection of the reflected light are merged in a common branch. Hence, illumination and detection are close together in front of the sample.

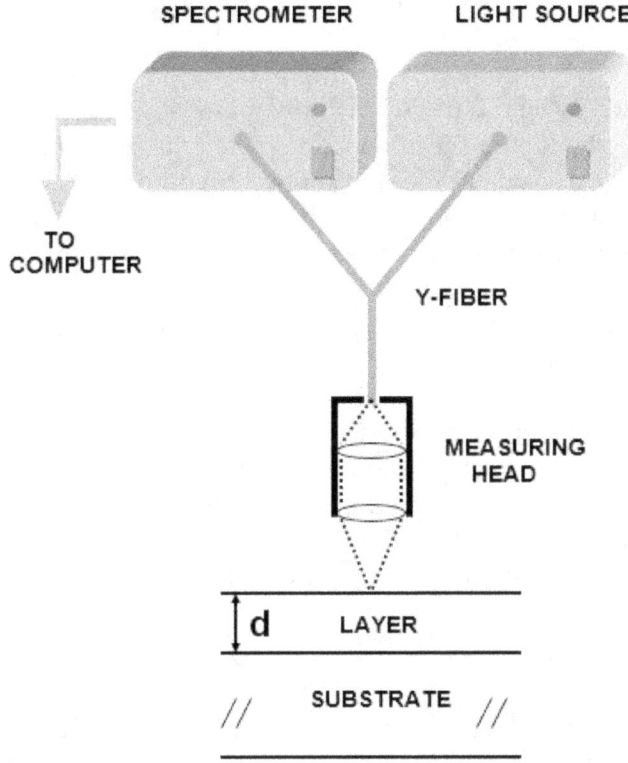

Fig. 4.2 Setup for reflectometric measurement with a miniaturized spectrometer

The measurement is carried out in the following way.

1.) Determine the dark current signal I_{dark}.
The detectors of miniaturized spectrometers are diode line arrays basing on two technologies
- CMOS (Complementary Metal Oxide Semiconductor) and
- CCD (Charged Coupled Device).

For the near infrared spectral region a third technology is used basing on InGaAs as detector material for each pixel.

All of them use the principle of the inner photoeffect where electrons get released by light but remain in the material. They change from the valence band into the conduction band and get stored in the potential well, a region in the semiconductor component, from which they are readout. The properties of the released electrons depend on various parameters: the absorption coefficient, the recombination time of the generated electron-hole pairs, the diffusion path, and the chemical and physical structure of the material above the photosensitive layer.

I_{dark} is a small electrical output of the detector array without incident light. It is caused by thermal generation of carriers in each pixels. It is therefore strongly correlated with the operation temperature. A further contribution to I_{dark} comes from the downward electronics in the AD-converter.

2.) Measurement of the light reflected at a reference sample

$$I_{ref}(\lambda) = R_{ref}(\lambda) \cdot S(\lambda) \cdot Sp(\lambda) + I_{dark} \tag{4.11}$$

where $S(\lambda)$ is the spectral distribution of the light source and $Sp(\lambda)$ is the spectrometer function that combines the spectral sensitivity of the detector, the transmission properties of the fibers, and some further spectrometer properties and their influence on the signal. The dark current signal is contained in the signal as almost constant value.

3.) Measurement of the light reflected at the sample

$$I(\lambda) = R_{sample}(\lambda) \cdot S(\lambda) \cdot Sp(\lambda) + I_{dark} \tag{4.12}$$

Also in this measurement the dark current signal is contained as almost constant contribution.

4.) Calculation of the ratio $I(\lambda)/I_{ref}(\lambda)$ and resolving for the sample reflectivity R_{sample}

$$R_{sample}(\lambda) = \frac{I(\lambda) - I_{dark}}{I_{ref}(\lambda) - I_{dark}} \cdot R_{ref}(\lambda).\tag{4.13}$$

This method presumes that the reflectivity R_{ref} of the reference sample is known, either as look-up table or by calculation from the optical constants and the thickness of the reference sample. The thickness of the reference sample is necessary for transparent or slightly absorbing reference materials, because then the reflection at the rear side of the reference sample is correctly taken into account.

Spectral reflectance measurement can be carried out without a mounted measuring head. Then, the common branch of the Y-fiber illuminates the sample and the light gets spread on the sample according to the numerical aperture NA of the fiber (typically NA = 0.22). Vice versa the detection fiber only collects the reflected light that enters its aperture. Therefore, the size of the detection spot is determined mainly by the core of the detection fiber, as from the widespread illumination spot only almost perpendicularly reflected light can enter the aperture of the detection fiber. Then, the measurement averages over the inhomogeneities in the film thickness in the area given by the core diameter of the detection fiber. On the other hand, this allows single measurements in short times (some tens of milliseconds).

Using a measuring head with reproduction scale 1:1 the incident light gets focused on illumination spots in the same order as the core diameter of the fiber. As the aperture angle of the focused beam corresponds to the aperture angle of the fiber the reflection measurement is practically not influenced from the polarization state of the incident beam and can be treated similar to normal incidence.

Spectral reflectance measurement with minituarized spectrometers is not free of errors. Sources of systematic errors are:

- The angle of incidence. This is already discussed above. As long as the angle of incidence is restricted on ±13° (NA = 0.22) deviation from the normal incidence the errors are negligible.
- An error source is that the detectors and amplifier circuits are not perfectly linear.

The determination of the complex refractive index n+iκ of a thin film from reflectance measurements is not independent of the thickness determination. All used methods simultaneously solve for thickness d *and* n+iκ, in spectral reflectance and ellipsometry measurements.
As only the reflectance gets measured at each wavelength provided by the spectrometer there is a big lack of information to retrieve both n and κ and even the film thickness d, because one has N measured data points but 2N + 1 unknown values. However, even if one additionally measures the transmittance, one has 2N measured data points but 2N +1 unknown values. Moreover, the measurement of transmittance is impossible when the substrate is opaque. The solution of this dilemma is to use dispersion models for n+iκ respectively the dielectric function $\varepsilon_1 + i \cdot \varepsilon_2$ as introduced in Chapter 3. Then, the number of unknowns gets strongly reduced on a few parameters for each used model plus the film thickness. The remaining task is to use a nonlinear fitting to obtain the parameters of the dispersion model and the film thickness by minimization of discrepancies between calculated and measured reflectance spectra.
The basic approach is to choose a figure-of-merit function that measures the agreement between the data and the model. Small values of the merit function represent close agreement. The parameters of the model are automatically adjusted so that the merit function achieves a minimum, yielding best-fit parameters. However, the merit function often has several local minima. Then, the task is not only to find a minimum but to find the global minimum. Otherwise, the determined parameters of the optical constants may be senseless.

Any fitting procedure should provide
- parameters,
- error estimates on the parameters, and
- a statistical measure of goodness-of-fit.

If each data point (x_i, y_i) of a set of N measured data has its own, known standard deviation σ_i then the maximum likelihood estimate of the model parameters $p_1, ..., p_M$ is obtained by minimizing the *quadratic deviation* χ^2

$$\chi^2 = \sum_{i=1}^{N}\left(\frac{y_i - f(x_i, p_1, ..., p_M)}{\sigma_i}\right)^2 \tag{4.14}$$

or the *Mean Squared Error* (MSE)

$$MSE = \sqrt{\frac{\chi^2}{2N-M}} \tag{4.15}$$

by iterative non-linear regression, with $f(x, p_1, ..., p_M)$ being the function that calculates the optical constants of the layer material as well as the reflectance of the layer stack system with the M parameters $p_1, , p_M$. A smaller χ^2 or MSE implies a better model fit to the data.

Various methods are available to find a minimum of χ^2. Unfortunately, there is no perfect algorithm. One method often used in nonlinear regression is the *Levenberg-Marquardt algorithm* [63, 64]. The Levenberg-Marquardt algorithm combines minimization of χ^2 with a Hessenberg matrix algorithm with the search for the steepest descent. For that purpose, it uses the first derivative of χ^2 versus each parameter p_i, $\partial\chi^2/\partial p_i$. This may be a long lasting method if the derivatives cannot be given as formula but must be calculated by variation of the parameter and calculating the differential quotient numerically. On the other hand, it allows fitting of many different parameters. Therefore, this algorithm is preferably used for determination of layer thickness, optical constants, and further parameters. It works well as long as the number of data points N in the measurement is sufficiently larger than the number of parameters M, i.e. N >> M. If χ^2 falls off or equals a certain value the calculation stops and the set of parameters with the best fit to the measured data is found. For more information we refer to the Numerical Recipes by Press *et al.* [65] and Appendix B.

The main problems with fitting reflectance spectra for determination of optical constants are however the adequate choice of the dispersion models and of the initial parameters for the unknowns. For solving these problems one has to check the literature thoroughly for all information on the material of the thin film. The initial estimates and the dispersion models must be derivable from the collected data. Otherwise, different results may be obtained for the optical constants when using different models. We will discuss this point again later in detail in Chapter 5.

4.2 Spectral Ellipsometric Measurement

Ellipsometry is a technique originally developed by Paul Drude to study polarized light reflected from solids coated with thin films and from liquid surfaces [66]. However just with the development of electronics and computers this over 125 years old technique became relevant because then it became possible to fit the experimental data to the physics-based first principles equations in a short time. Spectral ellipsometry (SE) can analyze complex structures such as multilayers, interface roughness, inhomogeneous layers, anisotropic layers and much more.

Azzam and Bashara [67] published in 1977 the book *Ellipsometry and Polarized Light* which has become the key source in ellipsometry. Further books that cover the theory of ellipsometry, fundamental principles, the instrumentation, and applications have been published later [68-73].

For an ellipsometric measurement a light source that provides unpolarized light and a polarizer are used to illuminate the sample with a light beam in an accurately known polarization state. Optionally, an optical retarder is placed between the polarizer and the sample. The direction of incidence includes an angle α with respect to the normal on the sample. Specular reflection of the beam from the sample surface leads to an emergent beam in an elliptical polarization state. It trespasses an analyzer and gets detected by an optical detector. Optionally, an optical retarder is placed between the sample and the analyzer. Usually, the analyzer is rotated (Rotating Analyzer Ellipsometry = RAE) to enable at least eight different measurements, but it may also be that the polarizer is rotated (Rotating Polarizer Ellipsometry = RPE) [29]. The principal setup for an ellipsometric measurement is sketched in Fig. 4.3.

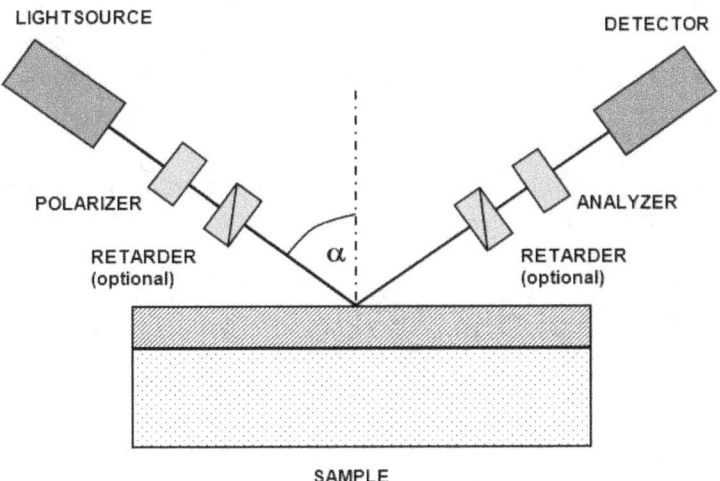

Fig. 4.3 Sketch of an ellipsometric thin film measurement

Ellipsometry measures the complex ratio ρ of the Fresnel reflection coefficient of the p-polarized and s-polarized component of the reflected light:

$$\rho = \frac{r_p(\alpha)}{r_s(\alpha)} = \frac{|r_p(\alpha)| \cdot \exp(i \cdot \Delta_p)}{|r_s(\alpha)| \cdot \exp(i \cdot \Delta_s)} = \sqrt{\frac{R_p}{R_s}} \cdot \exp(i\Delta) = \tan(\psi) \cdot \exp(i\Delta) \quad (4.16)$$

where tan(ψ) is the amplitude ratio and Δ is the phase shift of the p- and s- reflection coefficients. They are the *ellipsometric parameters* often also given as tan(ψ) and cos(Δ) or only as ψ and Δ.

Different measurement techniques of the polarization after reflection exist. For them other components like modulators or compensators can be added. Modern ellipsometer adjust all components automatically and calculate the ellipsometry parameters very fast.

Spectroscopic ellipsometry (SE) measures the change in polarization of light simultaneously at different wavelengths. The commercially available spectral range covers 150 nm to 33 μm but not in one single spectral range. It also allows the determination of the properties (thicknesses, complex refractive indices) of a layer stack. For that purpose, however, physical models for the layer stack and the optical constants are necessary. A regression analysis is used to get the best fit of calculated data to measured data.

If the sample is an ideal bulk, the real and imaginary parts of the pseudo complex dielectric function may be calculated from

$$\varepsilon_1 + i\varepsilon_2 = \sin^2(\alpha) \cdot \left[1 + \tan^2(\alpha) \cdot \left(\frac{1-\rho}{1+\rho}\right)\right]. \quad (4.17)$$

with the knowledge of the angle of incidence α and presuming the ambient to be air.

Ellipsometry has been extended to various measurement techniques. The major techniques are Phase or Polarization Modulation Ellipsometry (PME) [74-79], Two-Channel Phase Modulation Ellipsometry [80-82], and Variable Angle Spectroscopic Ellipsometry (VASE) [83-85].

Ellipsometric studies are generally carried out in the reflection mode rather than in the transmission mode, requiring either opaque substrates or substrates in which the backreflection is minimized or suppressed by different methods.

Ellipsometry is not free of errors. Sources of systematic errors are:

- Azimuthal alignment of optical elements

    Each optical element must be aligned to ensure a high quality ellipsometric measurement. The azimuthal alignment errors are typically in the order of 0.02° for each element. All alignment errors of the several optical elements add up and may result in a significant systematic error in the experiment.

- The angle of incidence

    The angle of incidence is a particularly difficult parameter to measure, and its error is hard to quantify. Generally, it is very hard to measure this quantity to better than $\approx 0.02°$. Additionally, the used light source is not perfectly collimated, so the sample is actually illuminated with a light beam with a distribution of angles of incidence.

- Calibration

    Spectroscopic ellipsometers that use retarders and compensators must calibrate the amount of phase shift as a function of wavelength. A further error source is that the detectors and amplifier circuits are not perfectly linear.

Ellipsometry is, however, less sensitive to surface roughness than many other methods. The reason is that the ratio of the polarized light intensities is measured rather than the absolute intensities.

The determination of the complex refractive index n+iκ of a thin film from ellipsometric measurements is also not independent of the thickness determination. All used methods simultaneously solve for thickness d *and* n+iκ.

Although the spectral ellipsometric measurement delivers two observables, ψ and Δ, we have to bear in mind that the two observables are not independent of each other. They belong to the same complex-valued quantity ρ. Therefore, we have the same dilemma as before in the reflectance measurement: we have N measured data points (ψ, Δ) but 2N + 1 unknown values. Again, one has to use dispersion models for $n+i\kappa$ respectively the dielectric function $\varepsilon_1 + i\cdot\varepsilon_2$ as introduced in Chapter 3 and to fit the calculated data set (ψ, Δ) to the measured. At this point of evaluation for optical constants and layer thickness there is no difference to the reflectance measurement described before in 4.2.1. For a review on modeling of spectroscopic ellipsometric data we refer to [86, 87].

5 Guidelines to the Practical Determination of Optical Constants

Once measured spectral reflectance R, R_p, R_s (and/or transmittance T, T_p, T_s) or the spectral dependence of ellipsometric parameters ψ and Δ for mostly a thin film on a substrate, the task is to retrieve the optical constants from these measurements. This is undoubted the main task we will have. Beyond this, however, there are further circumstances where it is meaningful to determine a convenient set of optical constants. All possible situations are listed below.

1) The investigated material is new, optical constants are not available. This is the main situation.
2) The material properties do change with composition. This situation may arise e.g. with ternary semiconductor compounds $A_xB_{1-x}Z$ where x is variable between 0 and 1.
3) The material properties depend on the preparation. For example, the stochiometric composition of silicon nitride is Si_3N_4, but can be variable Si_xN_y depending upon the preparation method of the silicon nitride film.
4) Data for the optical constants of the investigated material are sparsely available. That means that only for a few wavelengths we have tabulated optical constants. Any interpolation of the data on closer lying wavelengths results in jagged data.
5) Data for the optical constants of the investigated material are available but they are needed in another spectral range . Parametrization is then helpful to extend the spectral range of (n, κ) values.
6) Data are available but their use in calculations, e.g. in layerstack calculations or fits, lead to unsatisfactory results in parts or even in the whole spectral range.

Situations 1, 2, and 3 require first a good idea for the parametrization of the optical constants according to a model (physical or empiric) and then the fit of calculated spectral values to the measured spectral values. The results are the best-fit model parameters for the optical constants and the layer thickness. Situations 4, 5, and 6 require a parametrization of already existing data, and situation 6 furthermore a fit on measured data using the parametrization results as start parametrization.

In any case, we have

- to find the proper models for the layer material and the layer thickness,
- to adjust the initial values for the parameters, and
- to verify the results.

In this chapter we try to give tips and hints how to proceed in these three steps. For this we also use highly sophisticated software that allows for the calculation of reflectance spectra of layer stacks and ellipsometric parameters (ψ, Δ) and that allows for the calculation of optical constants according to different models as well as the simultaneous determination of film thickness and optical constants from reflectance measurement or spectral ellipsometry. For more information on this software please contact the author at michael.quinten@online.de.

5.1 How to Find Proper Models

The search for the proper models for the parametrization of optical constants is the most difficult part in the evaluation of the spectral measurements. This search often begins with the collection of all information that can be obtained about the material. For example, we consider the material ITO. ITO stands for indium-tin-oxide and is well-known as a transparent conductive oxide (TCO). Therefore, it is used in applications where a transparent electrode is required, as for example in OLEDs (organic light emitting diodes), in thin film photovoltaics, in flat panel displays, or in displays for mobile phones. ITO is a mixture of indium oxide In_2O_3 ($\approx$ 90%) and stannic oxide SnO_2 ($\approx$ 10%). In_2O_3 is a semi-conducting material with bandgap at 3.8 eV. The addition of SnO_2 generates impurities that are necessary to establish conductivity. The tin doped indium oxide In_2O_3:Sn finally has a bandgap at approximately 3.7 eV and a free carrier concentration of $N \cong 10^{21}$ cm^{-3}. However, as ITO is a mixture of two components the properties of ITO depend upon the actual composition. Therefore, optical constants of ITO from one vendor may be a bit different from the optical constants of another vendor, although the electrical properties and the transmittance are quite identical. What do we learn from this information about the optical constants? The first idea is that we can use a *Tauc-Lorentz oscillator* (see Chapter 3.1.3) that considers the bandgap. At photon energies lower than the bandgap the material must then become transparent. In addition, the contribution of the free charge carriers can be considered with the *Drude oscillator* model (see Chapter 3.1.2). The carrier concentration implies that we get weak absorption in the visible spectral region which increases when going to the near infrared and still lower photon energies. Combining both models we have a first attempt to the dielectric function of ITO. However, we still need information on the resonance frequency and the damping constant of the Tauc-Lorentz oscillator as well as the damping constant for the free charge carriers. The latter may be approximately obtained from the resistivity, as this damping constant as well as the carrier concentration enter the resistivity (resistivity = 1/DC-conductivity, DC-conductivity = Eq. (3.14)). Otherwise, we will try a damping constant of 1/100 - 1/10 of the plasma frequency. The oscillator resonance frequency may be obtained from a transmittance or absorbance spectrum. Otherwise, we will try a resonance frequency larger than the frequency corresponding to the band gap. In the example in Fig. 5.1 we used the following parameters for a model dielectric function of ITO: band gap E_{gap} = 3.7 eV, resonance frequency $\omega_0 = 10^{16}$ s^{-1} (E_0 = 6.582 eV), damping constant $\gamma = 10^{15}$ s^{-1} = $\omega_0/10$, plasma frequency $\omega_p = 2.5 \cdot 10^{15}$ s^{-1} (corresponding to a carrier concentration of $N = 1.964 \cdot 10^{21}$ cm^{-3}), and damping constant of the free carriers $\gamma_{fc} = 2.5 \cdot 10^{14}$ s^{-1} = $\omega_p/10$. With these parameters we calculated the dielectric function, respectively the complex refractive index in the wavelength range 200 nm to 1200 nm and compared in this approach with data from literature [88-90].

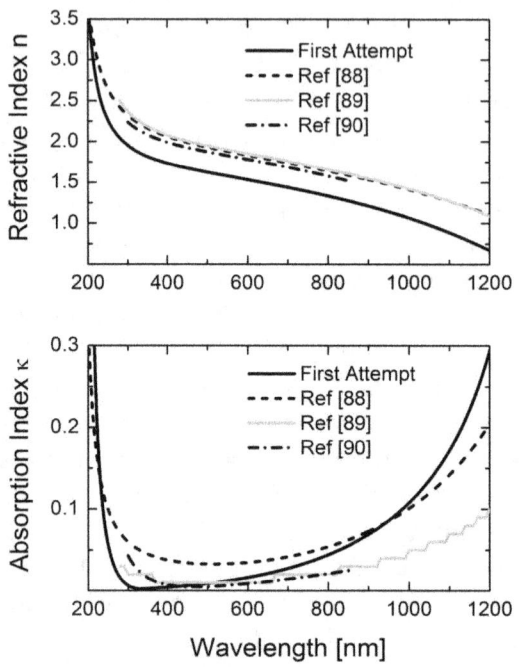

Fig. 5.1 Optical constants of ITO according to the model Tauc-Lorentz oscillator plus Drude oscillator in comparison to optical constants from literature [88-90]

Obviously, this first approach is pretty close to the data in the given references. Hence, the above parameters are already well-suited as initial values for a fit of optical constants of ITO.

If we put now a thin film of ITO on a glass we do not only measure the reflectance of the interface ITO-air but we get film thickness interferences that modify the reflectance spectrum in a characteristic way. This is demonstrated in Fig. 5.2 using the optical constants from [88].

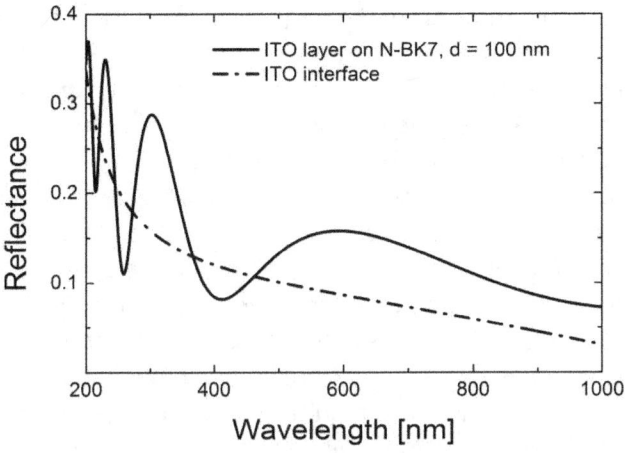

Fig. 5.2 The reflectance of an ITO film with thickness d = 100 nm and optical constants from [88] in comparison to the reflectance of the interface ITO-air

The reflectance of the ITO film is modified by oscillations that are the result of the interference of the light reflected at the interface air-film and reflected at the interface film-substrate. For explanation and a brief derivation of the reflectance of a thin film on a substrate we refer to Chapter 2.3. Here, we only repeat the result

$$R(\lambda,d) = \frac{R_{01}(\lambda) + R_{12}(\lambda) + 2\sqrt{R_{01}(\lambda) \cdot R_{12}(\lambda)} \cdot \cos\left(\frac{4\pi}{\lambda} n_1(\lambda) \cdot d\right)}{1 + R_{01}(\lambda) \cdot R_{12}(\lambda) + 2\sqrt{R_{01}(\lambda) \cdot R_{12}(\lambda)} \cdot \cos\left(\frac{4\pi}{\lambda} n_1(\lambda) \cdot d\right)}.$$

(5.1)

R_{01} is the reflectance of the interface air-film and R_{12} the reflectance of the interface film-substrate.

These oscillations render the determination of the optical constants more difficult twice. First, they introduce a periodic structure in the spectrum and hence may superpose structures that originate from the dispersion of the optical constants. Second, the calculation of the reflectance must be extended on the calculation of the reflectance of (5.1) or even more complex relations in the case of a layer stack to consider also the unknown film thickness. The best way to a reasonable solution of this multivariate problem is to proceed as follows.

First load the measured reflectance spectrum into a highly sophisticated software that can model optical constants as well as can calculate reflectance spectra of layer stacks. Second, define a first attempt to the optical constants of the film material from your search. Third, give an approximate value for the thickness of the layer. It can be obtained from other independent measurements or from your expectations. After this the software should show you the resulting reflectance in comparison to the measured reflectance. Now, it is time to vary the parameters of the optical constants and the film thickness manually and to observe whether the agreement of calculation and measurement becomes better or worse. For this the software optimally recalculates and displays the new reflectance spectrum directly and also displays the value of the quadratic deviation χ^2. If the agreement is fairly good the fit can be started. To improve the agreement also during the manual fit the software should be able to add or to remove models to/from your attempt.

In the following, we demonstrate this procedure using such a highly sophisticated software, named MQNandK. For more information on this software, please contact the author at michael.quinten@online.de.
After having started this software, we select *Fit R, T* to begin with the fit. Then, a dialogue and an empty graphics window appear simultaneously in the main window as shown in Fig. 5.3.

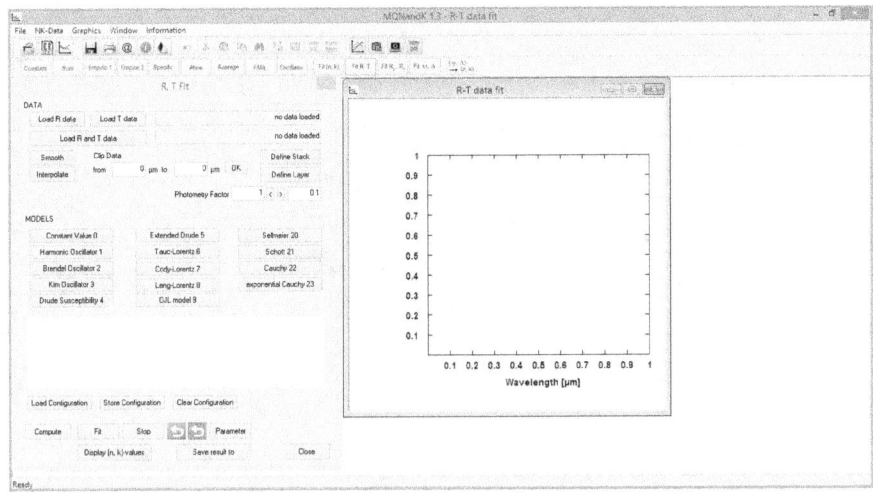

Fig. 5.3 Start screen of the menu *Fit R, T* of the software MQNandK

Looking first at the dialogue in Fig. 5.4 (next page). Load the measured data in the group DATA. In our example only reflectance data get read in using the button *Load R data*. It is also possible to read in and fit only transmittance data, or to read in and fit reflectance *and* transmittance data. After this the dialogue for the definition of the layerstack (Fig. 5.5, next page) automatically opens. It also opens when you strike the button *Define Layerstack*. Here, the composition of the stack on which the reflectance and transmittance were measured is given and an approximate thickness of the unknown layer.

After confirming this dialogue with *OK* or *Close* the graphics window gets refreshed and shows the measured reflectance data as black curve. In the example, we used the calculated reflectance spectrum of an ITO layer with thickness d = 100 nm on a N-BK7 glass substrate already shown in Fig. 5.2 as "measured" reflectance spectrum.

Fig. 5.4 Dialogue of the menu *R,T Fit*

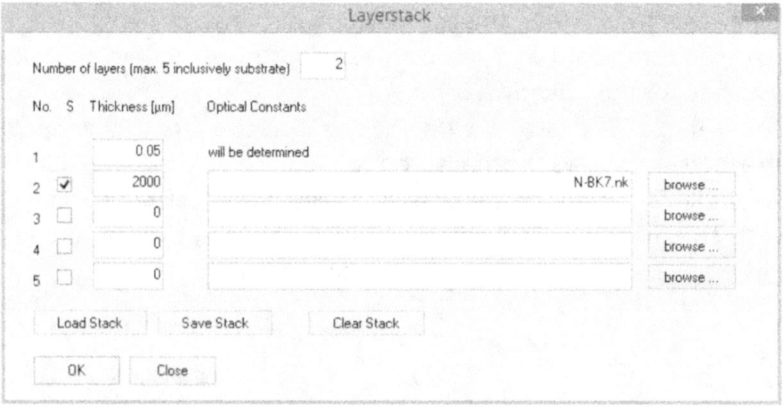

Fig. 5.5 Dialogue for definition of the layer stack

Below the graphics window a further dialogue (Fig. 5.6) opens where the parameters of the layer can be fixed. Give here an initial value for the thickness.

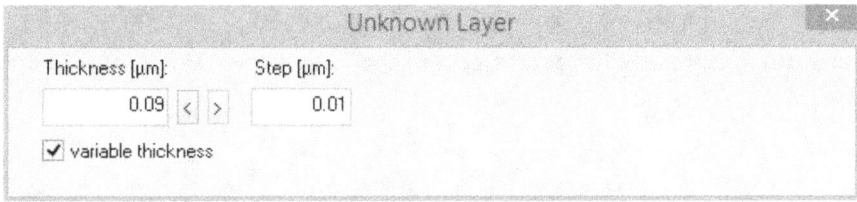

Fig. 5.6 Dialogue for the thickness of the unknown layer

To parametrize the optical constants use the buttons in the group MO-DELS. In our example of an ITO layer we try the models *Tauc-Lorentz* (model number 6) and *Drude Susceptibility* (model number 4). When striking the corresponding button a new dialogue opens for each model where the model parameters can be given as illustrated in Fig. 5.7.

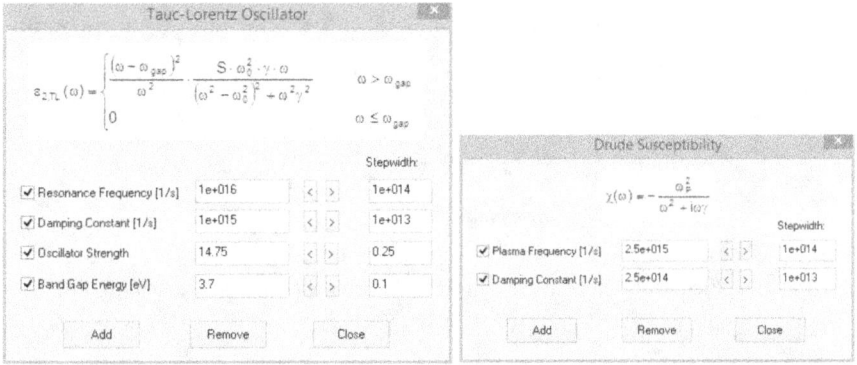

Fig. 5.7 Dialogues of the models Tauc-Lorentz oscillator and Drude Susceptibility

We insert the parameters used in our first attempt for the model dielectric function of ITO: band gap E_{gap} = 3.7 eV, resonance frequency ω_0 = 10^{16} s^{-1}, damping constant γ = 10^{15} s^{-1} for the Tauc-Lorentz oscillator and plasma frequency ω_p = 2.5·10^{15} s^{-1} and damping constant of the free

carriers $\gamma_{fc} = 2.5 \cdot 10^{14}$ s^{-1} for the Drude susceptibility. The parameters get written in the list below the model buttons (see Fig. 5.4).
In the graphics window (Fig. 5.8) now the green curve shows the calculated reflectance. The rather good coincidence of measured and calculated data is obtained by adjusting manually the film thickness to 0.09 µm and the oscillator strength of the Tauc-Lorentz oscillator to S = 14.75.

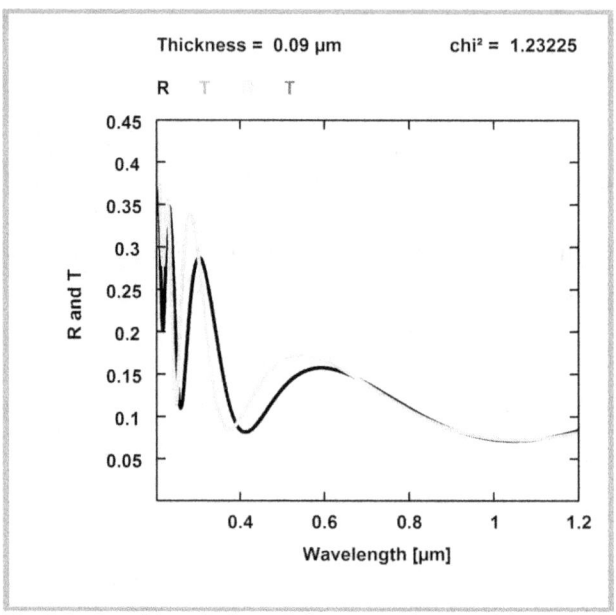

Fig. 5.8 Graphics window of the menu Fit R, T with measured and calculated data

Measured and calculated curves deviate with a *quadratic deviation* χ^2:

$$\chi^2 = \frac{1}{N-M} \sum_n^N \frac{\left(R(\lambda_n) - R_{calc}(\lambda_n, p_1, \ldots, p_M)\right)^2}{\sigma(\lambda_n)^2} \tag{5.2}$$

Note, that in the software MQNandK the standard deviation $\sigma(\lambda_n)$ is always set to 1 for each measured point. Moreover, χ^2 is not normalised to N-M.

The task is to minimize χ^2. If χ^2 is minimal the optimum parameters are found. The quadratic deviation can be minimized either manually or automatically. Manually means that the parameters are changed stepwise by the user until a minimum of χ^2 is almost approached. This method may be insufficient, but should be used absolutely before an automatic fit to approach the best initial values of the fit. The reason is that the automatic fit can only find meaningful results if the initial values are in the vicinity of the final results. A big distance of one or more parameters can result in only a local minimum of χ^2 instead of the global minimum. The automatic fit gets started with striking the button *Fit* in the dialogue in Fig. 5.4. The fit is based on the Levenberg-Marquardt method. After a successful fit the graphics window in Fig. 5.9 and the configuration list in the dialogue are updated with the calculated model parameters.

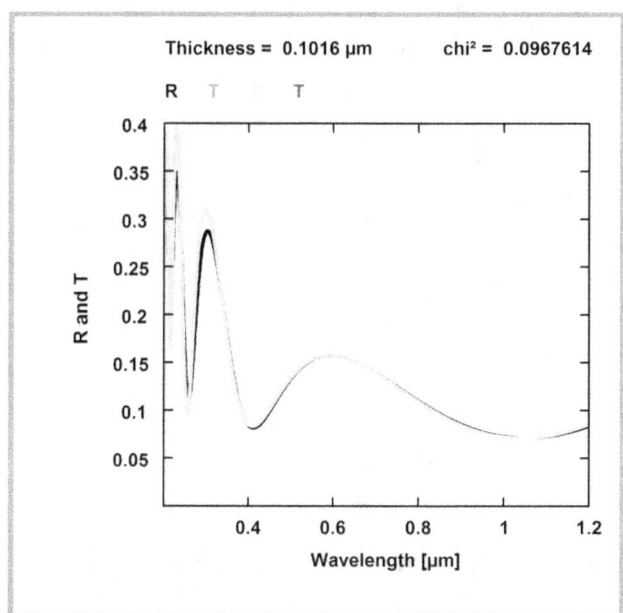

Fig. 5.9 Final fit result

The obtained thickness of d = 101.6 nm matches well to the actual thickness of d = 100 nm. The obtained optical constants are compared with the optical constants from [88] as these were used in the calculation of the "measured" spectrum. The comparison in Fig. 5.10 shows that we arrived in a pretty good agreement with the refractive index of Ref. [88]. The absorption index shows small deviations. The obtained parameters are oscillator strength S = 12.7231, band gap E_{gap} = 3.66665 eV, resonance frequency ω_0 = $1.05395 \cdot 10^{16}$ s^{-1}, and damping constant γ = $5.38679 \cdot 10^{14}$ s^{-1} for the Tauc-Lorentz oscillator and plasma frequency ω_p = $2.58724 \cdot 10^{15}$ s^{-1} and damping constant of the free carriers γ_{fc} = $2.95119 \cdot 10^{14}$ s^{-1} for the Drude susceptibility.

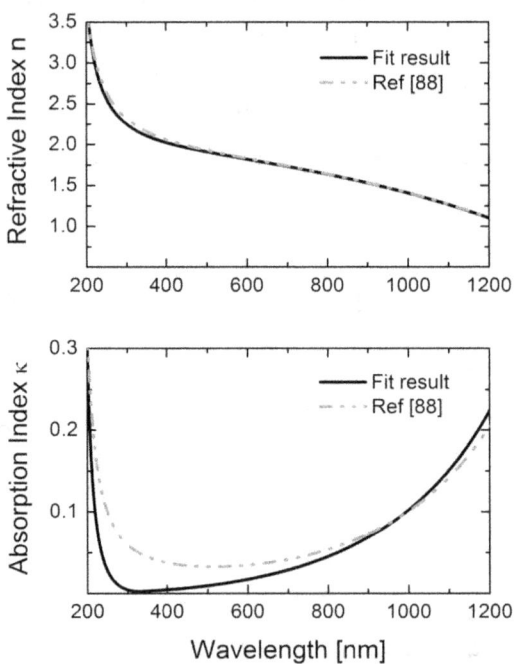

Fig. 5.10 Comparison of the optical constants obtained in the above fit with the optical constants from [88]

The demonstrated example worked well and we arrived in a pretty good result within a few steps. In many cases however, mainly the search for the best model for the optical constants is more complex and time-consuming. In that cases, you must try several model configurations to determine the most appropriate number of models and the most appropriate models.

A problem may arise with the film thickness. If the film thickness becomes large we can observe rapid oscillations of the reflectance as interference pattern. These rapid oscillations make the fit ambiguous since they may overlay spectral features resulting from the optical constants rather than from thickness interferences. For a proper fit the film thickness should therefore range from 100 nm to 500 nm. Nevertheless, for some film materials that are predominantly transparent in the considered wavelength range also for thicker films the determination of optical constants can be successful. This is demonstrated in the following example.

We consider a photoresist layer on silicon of approximate thickness of 4 μm. We first think again of a proper model for the optical constants. As such photoresists are commonly illuminated with blue/violet or ultraviolet light to modify the consistency of the material in certain regions, it must be absorbing at such short wavelengths. Hence, harmonic oscillators and broadened harmonic oscillators can be used for modeling the optical constants. Unfortunately, the current thickness of the film did not allow to determine resonance frequencies from the measured reflectance spectrum, as it is as large as the absorption already destroys characteristic features in the reflectance spectrum below 330 nm wavelength. This can be seen in Fig. 5.11 where the measured spectrum of this photoresist film is shown. On the other hand, it can clearly be recognized that the absorption constricts the rapid oscillations when going from long to short wavelengths. Therefore, it seems appropriate to use a much simpler empiric model for the optical constants of the photoresist, the empiric model *exponential Cauchy* (see Chapter 3.2). With this model we approximate the dispersion of n quite reasonable and approximate the decreasing absorption from short to long wavelengths with an exponential decrease of κ. The resulting parametrized optical constants n and κ are given as

$$n = 1.4943 + \frac{3.97081 \cdot 10^3}{\lambda^2} + \frac{1.34705 \cdot 10^9}{\lambda^4} \qquad (5.3)$$

$$\kappa = 0.171018 \cdot \exp\left(9.74722 \cdot \left(\frac{1239.841875}{\lambda}\right) - 3.88003\right) \qquad (5.4)$$

with the wavelength λ in nm.

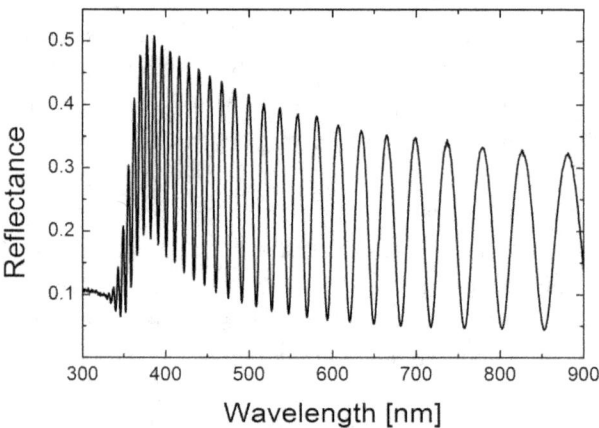

Fig. 5.11 Reflectance spectrum of a photoresist layer with approximate thickness d = 4 μm on a silicon wafer

In the next example we demonstrate how to proceed in the case that optical constants (n, κ) are already available, but only at distinct wavelengths sometimes being a long way away from each other. Interpolation of the data between these wavelengths often lead to jagged data. A parametrization can be helpful to get a smooth set of optical constants. We consider here the optical constants of titanium dioxide (rutile modification) for the ordinary ray tabulated in the *Handbook of*

Optical Constants I [1]. They are shown in Fig. 5.12 as grey dots that are connected with a thin straight line between two adjacent points. As one can recognize the data are widespread so that an interpolation between the data would lead to jagged data. Partly, this can be seen already from the thin connecting lines between the data.

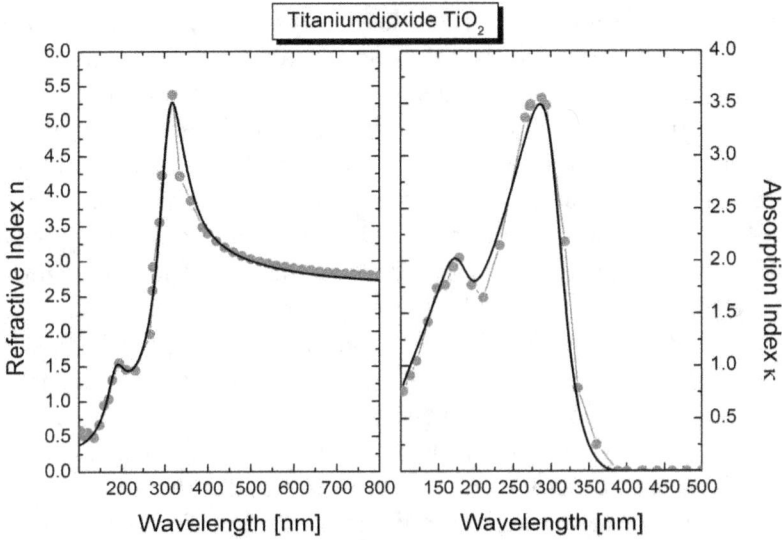

Fig. 5.12 Optical constants of rutile for the ordinary ray: grey dots and thin grey lines: as tabulated in [1], black line: approximation with two Tauc-Lorentz oscillators

We tried to get smoother data by parametrization. For this, again, we must first think of suited models for the optical constants.

The relatively abrupt vanishing of the absorption index at around 390 nm wavelength indicates that TiO_2 exhibits a bandgap around this wavelength. Therefore, a *Tauc-Lorentz oscillator* (see Chapter 3.1.3) that considers the bandgap seems to be the appropriate choice. A second oscillator at around 180 nm must also be a Tauc-Lorentz oscillator with the same bandgap, since any other possible oscillators (Lorentz oscillator, Brendel, Kim) would lead to nonzero absorption coefficient at wavelengths larger than 390 nm. Our approximation succeeded with two Tauc-Lorentz oscillators with a band gap E_{gap} = 3.19 eV, resonance frequencies ω_1 = 6.12648·10^{15} s^{-1} and ω_2 = 1.03391·10^{16} s^{-1} (E_1 = 4.03252 eV, E_2 = 6.80532 eV), and damping constants γ_1 = 1.10502·10^{15} s^{-1} and γ_2 = 2.79079·10^{15} s^{-1}.

Finally, we demonstrate in the last example how to proceed when available data do not lead to satisfying results when using them in fits on measured reflectance or transmittance data of a thin film. The example is from thin film thickness determination of SuperYellow films.
SuperYellow is a high performance OLED material from Merck KGaA & Co. KG, Germany, that can be processed by spin-coating and printing. Thin films of this material on a glass sheet are yellow-colored. However, the color of the thin film does not only depend upon the absorbance of SuperYellow but also on the film thickness. To determine the film thickness from reflectance measurement commonly a fit of a calculated reflectance spectrum on the measured spectrum is carried out. For this, the optical constants of SuperYellow must be known in the spectral range. In Fig. 5.13 we comprise measured reflectance spectra of SuperYellow films of various thickness and layer thickness fits using optical constants determined at the Institute of Physical Chemistry, University of Cologne, Germany using ellipsometry [91]. The fits were done with another highly sophisticated software for calculation and fit of layer stacks called MQLayer.
All calculated spectra show a lack of agreement particularly in the short wavelength range. For improvement of the fit we try to determine the optical constants of SuperYellow and the film thickness simultaneously for each examined sample. In contrast to the examples before we already have a pretty good set of optical constants for SuperYellow. Therefore, we first try to parametrize these optical constants to obtain start parameters for our fit on the measured reflectance data. For this, we use again the menu *Fit (n,k)* of the software MQNandK.

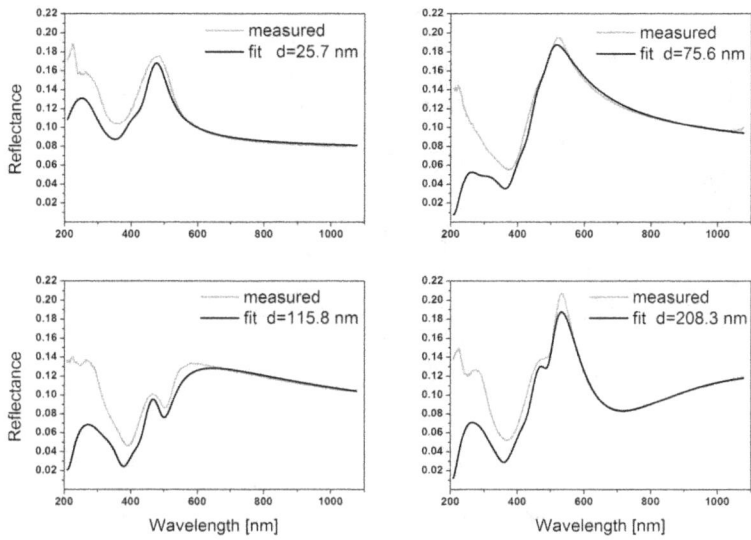

Fig. 5.13 Measured reflectance spectra and layer thickness fits of SuperYellow thin films on N-BK7, using optical constants from [91]

Looking at the refractive index n and the absorption index κ obtained from [91] in Fig. 5.14 we can recognize that the absorption index κ exhibits two clear maxima and in addition a third maximum close to the second maximum. This indicates that we can try to approximate the optical constants either with three harmonic oscillators or three broadened harmonic oscillators (Brendel, or Kim). Typical features of a band gap cannot be recognized, for what we exclude Tauc-Lorentz oscillators and related oscillators. In fact, we succeed in a rather good agreement with the optical constants from [91] using three Kim oscillators. This fit is also shown in Fig. 5.14.

Now, we have a basis for the determination of the optical constants and the film thickness from the measured reflectance data in Fig. 5.13. Using again the menu *Fit R, T* in MQNandK we determined optical constants (ε_1, ε_2) of SuperYellow for each investigated sample and averaged them.

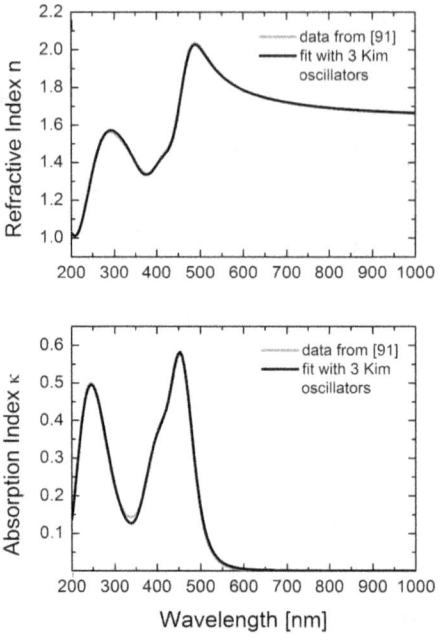

Fig. 5.14 Optical constants of SuperYellow from [91] (grey lines) and fit with three Kim oscillators (black lines)

Note that only dielectric functions can be averaged, not refractive indices. The corresponding (n, κ) - data of this average dielectric function were finally inserted again in a thickness fit with MQLayer. The agreement of the calculated curves with the measured reflectances spectra in Fig. 5.15 has been remarkably improved by the averaged optical constants.

While the film thickness varies little in comparison to the fits in Fig. 5.15, the main improvement of the fits results from the properly determined optical constants of SuperYellow. This can be followed from the quadratic deviation χ^2. The values for χ^2 and for the obtained layer thickness d are comprised in Table 5.1.

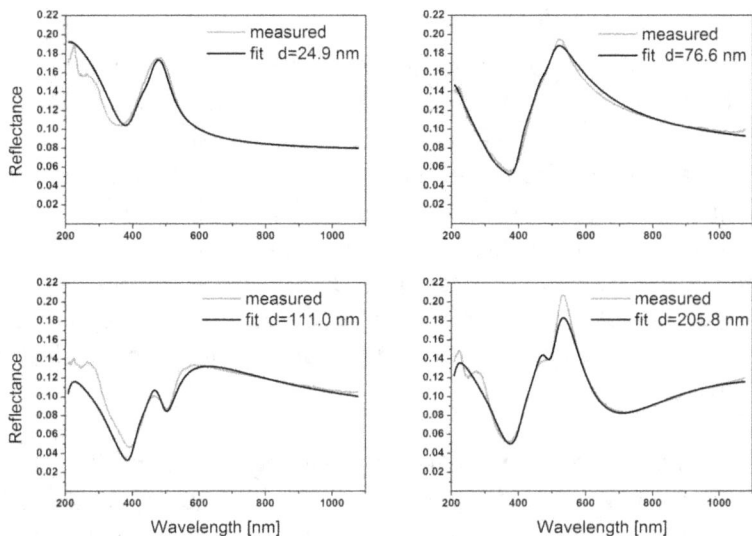

Fig. 5.15 Measured reflectance spectra and layer thickness fits of SuperYellow thin films on N-BK7, using optical constants from a fit with three Kim oscillators

Table 5.1 Film thickness and quadratic deviation from fit of thin film reflectance spectra of SuperYellow.

Sample	with optical constants from [91]	with optical constants from parametrization with three Kim oscillators
1	d = 25.7 nm, χ^2 = 0.57518	d = 24.9 nm, χ^2 = 0.13026
2	d = 75.6 nm, χ^2 = 1.6322	d = 76.6 nm, χ^2 = 0.027484
3	d = 115.8 nm, χ^2 = 1.87088	d = 111.0 nm, χ^2 = 0.28433
4	d = 208.3 nm, χ^2 = 1.70716	d = 205.8 nm, χ^2 = 0.07761

5.2 How to Adjust Initial Values for the Fit

In Sect. 5.1 we gave examples for the determination of optical constants. The attention was to find the proper model. Once found the proper model the fit can still fail because the initial values for the model parameters are too far from the final result. Then, the fit will run into a local minimum of the quadratic deviation χ^2 instead to find the global minimum. To prevent such a scenario the initial values for the fit must be manually adjusted. This process gets demonstrated in the following with the example of an ITO film on glass as in Sect. 5.1. In a sequence of pictures we show how the calculated fit curve approaches or departs the measured curve.

We use again the models *Tauc-Lorentz* with band gap $E_{gap} = 3.7$ eV, resonance frequency $\omega_0 = 10^{16}$ s^{-1}, damping constant $\gamma = 10^{15}$ s^{-1}, and oscillator strength S = 5 as initial values and a *Drude Susceptibility* with plasma frequency $\omega_p = 2.5 \cdot 10^{15}$ s^{-1} and damping constant of the free carriers $\gamma_{fc} = 2.5 \cdot 10^{14}$ s^{-1}. The film thickness is assumed to amount to 100 nm.

The resulting calculated reflectance spectrum is shown in Fig. 5.16a together with the measured one. The agreement is quite bad with a quadratic deviation of $\chi^2 = 8.89478$. Now we vary the oscillator strength S of the Tauc-Lorentz oscillator. The initial value of S = 5 in the first attempt is obviously too low. With S = 10, 15, and 20 in Fig. 5.16b, c, and d we converge to a reasonable starting point for an automatic fit with the result S = 15 as initial value. For a larger oscillator strength χ^2 becomes larger again.

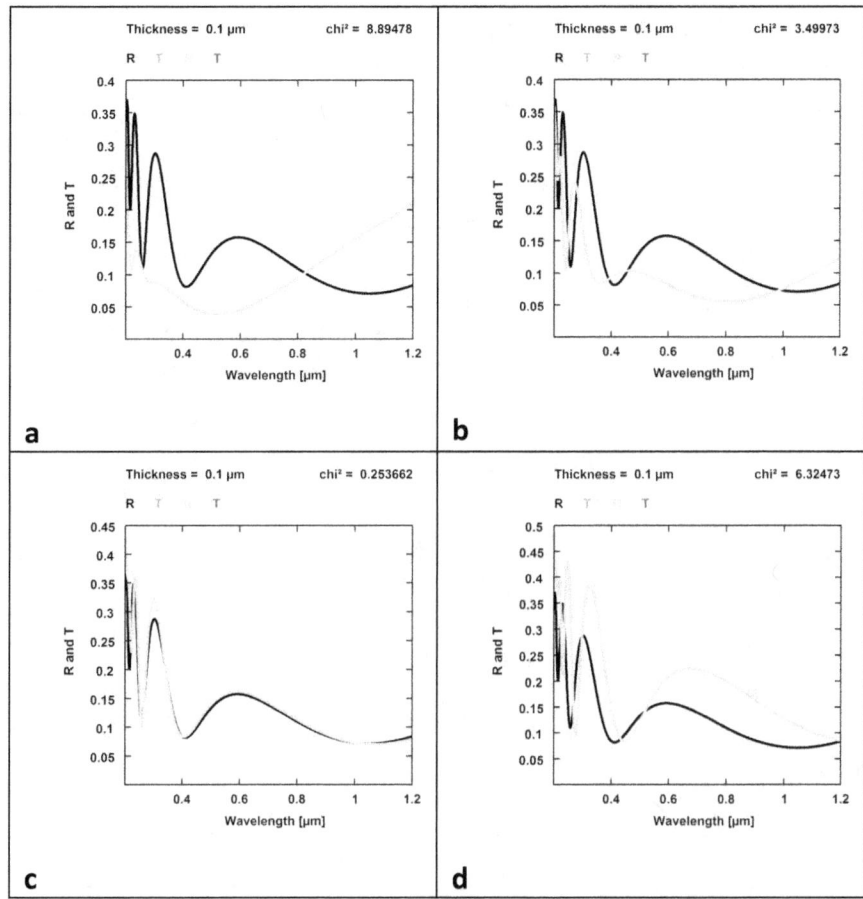

Fig. 5.16 Reflectance spectrum of an ITO-layer with d = 100 nm on N-BK7 and the fit using the a Tauc-Lorentz oscillator with varying oscillator strength S and a Drude susceptibility: a.) S = 5, b.) S = 10, c.) S = 15, d.) S = 20

In a second sequence we demonstrate the convergence by varying the film thickness. In Fig. 5.17a we start with an oscillator strength of already S = 15, but with d = 0.05 μm and vary then the film thickness to d = 0.075, 0.1, and 0.125 in Fig. 5.17b, c, and d. A rather satisfactory result is obtained with d = 0.1 µm.

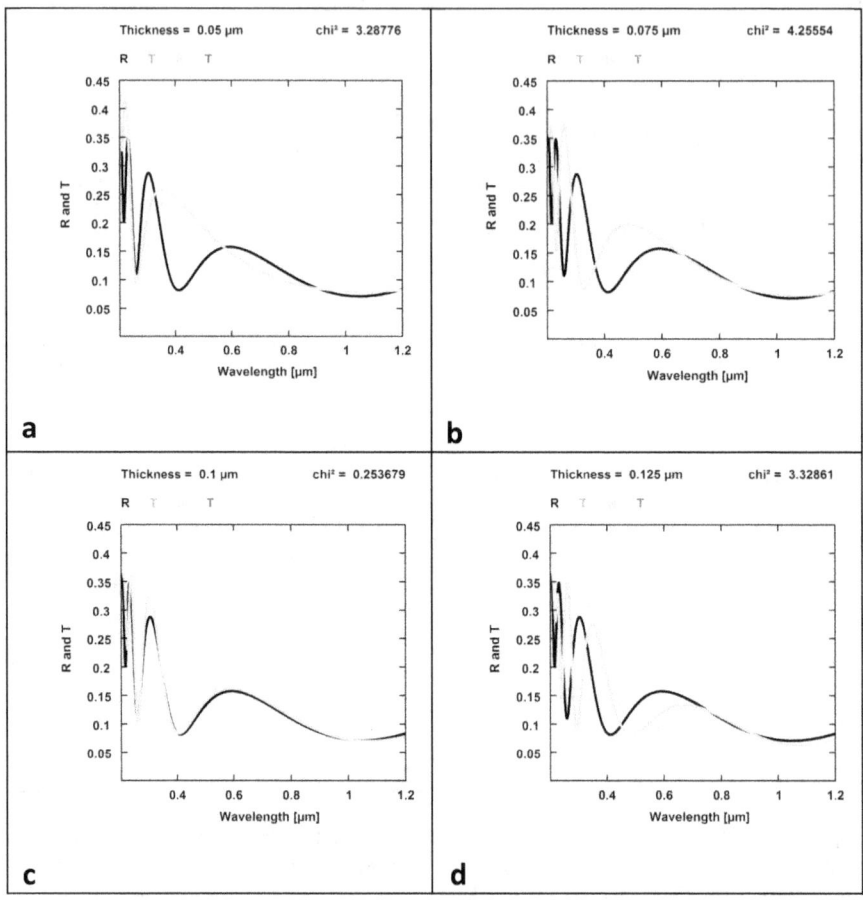

Fig. 5.17 Reflectance spectrum of an ITO-layer with varying layer thickness d on N-BK7 and the fit using the a Tauc-Lorentz oscillator and a Drude susceptibility: a.) d = 50 nm, b.) d = 75 nm, c.) d = 100 nm, d.) d = 125 nm

In a similar way all other parameters of the model for the optical constants can be manually adjusted to improve the fit. This is continued until the fit curve deviates little from the measured data. Then, an automatic fit can be started. The key to a pretty good set of initial values for an automatic fit is the visual examination of the fit curve in comparison to the measured curve. This is also helpful for improvement of the fit by either adding further oscillators or to remove oscillators if necessary.

5.3 How to Verify the Results

Any kind of fit procedure yields parameters that minimise the quadratic deviation between the measured data and calculated data. The obtained parameters are therefore mathematically meaningful but may be physically meaningless. The problem is to check whether they are also physically meaningful. Here, we give some tips how to verify the parameters. Nevertheless, sometimes the parameters may still be meaningless although they appear to be meaningful.

The most critical model parameters are the constant susceptibility term, the oscillator resonance frequencies and the plasma frequency of free charge carriers. To prove whether they are meaningful, one should first recalculate the wavelengths of the used spectral range into circular frequencies and photon energies. For this use the relations given in Chapter 1 (Eqs. (1.4) - (1.6)). If for example the wavelength range is from 400 nm to 1000 nm, the corresponding frequency range is from $4.7 \cdot 10^{15}$ s^{-1} (400 nm) to $1.884 \cdot 10^{15}$ s^{-1} (1000 nm) and the corresponding photon energies are from 3.1 eV (400 nm) to 1.24 eV (1000 nm). Therefore, resonance frequencies of resonators *in* the considered wavelength should be in the order of 2 - $5 \cdot 10^{15}$ s^{-1}. Resonators *outside* the considered wavelength range cannot have resonance frequencies too far away from this frequency range, otherwise their contribution to the optical constants in this frequency or wavelength range becomes negligible. A reasonable estimation are frequencies 4 times larger than the largest frequency (here $5 \cdot 10^{15}$ s^{-1}) and 4 times smaller than the smallest frequency (here $2 \cdot 10^{15}$ s^{-1}). The validity of the plasma frequency can be checked against the plasma frequency of good metals. No material can have higher plasma frequencies as good metals (Cu, Ag, Au, Al). They lie in the range of 1.4 - 2.4·10^{16} s^{-1}. Larger plasma frequencies are doubtful. On the other hand plasma frequencies less than 10^{14} s^{-1} are also meaningless considering the contribution of free charge carriers to the optical constants in the considered wavelengh range. Last, the constant susceptibility can be a critical value. As it theoretically is the long wavelength remainder of all resonances at shorter wavelengths (in our example: from all excitations in the X-ray region) it cannot amount to very high values. In our example values above 10, but also negative values below -5 are questionable.

These considerations can be extended to the other parameters, particularly on band gap energies. Damping constants are less critical.

6 Appendices

6.1 Appendix A: Numerics With Complex Numbers

Complex numbers are useful abstract quantities that can be used in calculations and result in physically meaningful solutions. Complex numbers have been introduced to allow for solutions of certain equations that have no real solution. For example, the quadratic equation $x^2 + 1 = 0$ has no solution in the field of real numbers. Complex numbers are a solution to this problem. The complex numbers are the field $\mathbb{C}$ of numbers of the form $z = x + i \cdot y$, where $x, y \in \mathbb{R}$, the field of real numbers, and i is the *imaginary unit* $i = \sqrt{-1}$. They extend the idea of the one-dimensional number line to the two-dimensional complex plane using the number line for the *real part* (x-values) and adding a vertical axis for the *imaginary part* (y-values). The graphical representation of the complex number z in the complex plane is sketched in Fig. 6.1.

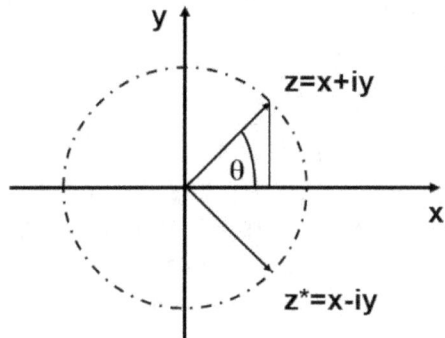

Fig. 6.1 Graphical representation of complex numbers

If $z = x + i \cdot y$ is a complex number, then x is called *real part* of z, i.e. Re(z) = x. Analogously, Im(z) = y is called the *imaginary part* of z.

If $z = x + i \cdot y$ is a complex number, then $z^* = x - i \cdot y$ is the complex number that lies in the conjugated plane, and is therefore called *complex conjugate number*.

In the field $\mathbb{C}$ of complex numbers there are two operations defined: "+", meaning addition, and "•", meaning multiplication.

Addition

The addition of the complex numbers z_1 and z_2 is defined as

$$z_1 + z_2 = (x_1 + i \cdot y_1) + (x_2 + i \cdot y_2) = (x_1 + x_2) + i \cdot (y_1 + y_2). \tag{6.1}$$

The addition is commutative, i.e. $z_1 + z_2 = z_2 + z_1$.
The neutral element of the addition is $n(+) = 0 + i \cdot 0 = 0$.
For the inverse element of the addition inv^+ it is $z + inv^+(z) = n(+)$, resulting in $inv^+(z) = -z$.

Multiplication

The multiplication of two complex numbers z_1 and z_2 is defined as

$$z_1 \cdot z_2 = (x_1 + i \cdot y_1) \cdot (x_2 + i \cdot y_2) = (x_1 \cdot x_2 - y_1 \cdot y_2) + i \cdot (x_1 \cdot y_2 - x_2 \cdot y_1) \tag{6.2}$$

The multiplication is commutative, i.e. $z_1 \cdot z_2 = z_1 \cdot z_2$.
The neutral element of the multiplication is $n(\bullet) = 1 + i \cdot 0 = 1$.
For the inverse element of the multiplication $inv^\bullet$ it is $z \cdot inv^\bullet(z) = n(\bullet)$, resulting in $inv^\bullet(z) = 1/z$ for all complex numbers $\neq 0$. With the help of the complex conjugate number z^* it can be expressed as

$$inv^\bullet(z) = 1/z = z^*/(z \cdot z^*) \tag{6.3}$$

Modulus

The modulus of a complex number $z = x + i \cdot y$ corresponds to the length of the pointer in Fig. 6.1. It is the hypotenuse of the triangle formed by the real part x and the imaginary part y as legs of a right-angled triangle. Therefore, the modulus |z| follows as

$$|z| = \sqrt{x^2 + y^2}. \qquad (6.4)$$

For all complex numbers with the same modulus the corresponding pointer ends on the dash-dotted circle in Fig. 6.1. From this graphical representation we can deduce that

$$z = x + i \cdot y = |z| \cdot (\cos(\theta) + i \cdot \sin(\theta)) = |z| \cdot \exp(i\theta) \qquad (6.5)$$

with $\theta = \arg(z) = \tan^{-1}(y/x)$ being the *argument* of z. This is the polar representation of a complex number. The analytical identity $\cos(\theta) + i \cdot \sin(\theta) = \exp(i \cdot \theta)$ allows for the application of power laws when calculating with complex numbers. The multiplication of z with its complex conjugate number z* yields $|z|^2 = z \cdot z^*$.

Division
The division of two complex numbers z_1/z_2 can be reformulated into a multiplication of two complex numbers z_1 and $(1/z_2) = \text{inv}^\bullet(z_2)$. Hence, the division is defined as

$$\frac{z_1}{z_2} = z_1 \cdot \text{inv}^\bullet(z_2) = \frac{z_1 \cdot z_2^*}{z_2 z_2^*} \qquad (6.6)$$

Power n
To calculate z^n with n being a real number $n \in \mathbb{R}$, the polar representation of a complex number (7.5) is useful. Then

$$z^n = |z|^n \cdot (\cos(n\theta) + i \cdot \sin(n\theta)) = |z|^n \cdot \exp(in\theta) \qquad (6.7)$$

For n ∈ N, i.e. a positive integer number, we can derive that z^n is given by

$$z^n = \left[x^n - \binom{n}{2}x^{n-2}y^2 + \binom{n}{4}x^{n-4}y^4 - \ldots \right] + i \cdot \left[\binom{n}{1}x^{n-1}y - \binom{n}{3}x^{n-3}y^3 + \ldots \right].$$

(6.8)

Logarithm
The natural logarithm log(z) of a complex number z can easily be calculated using again the polar representation (6.5):

$$\log(z) = \log(|z| \cdot \exp(i\theta)) = \log(|z|) + i \cdot \theta.$$

(6.9)

Exponentiation
For the complex exponentiation $z_1^{z_2}$ we can use the exponential function:

$$z_1^{z_2} = (\exp(\log(z_1)))^{z_2} = \exp(z_2 \cdot (\log(|z_1|) + i\theta_1)).$$

(6.10)

Trigonometric functions
The trigonometric functions sin(z) and cos(z) can be calculated using the analytical identity cos(z) + i·sin(z) = exp(iz). Then

$$\sin(z) = \frac{\exp(+iz) - \exp(-iz)}{2 \cdot i} = \sin(x) \cdot \cosh(y) + i \cdot \cos(x) \cdot \sinh(y) \quad (6.11)$$

$$\cos(z) = \frac{\exp(+iz) + \exp(-iz)}{2} = \cos(x) \cdot \cosh(y) - i \cdot \sin(x) \cdot \sinh(y) \quad (6.12)$$

6.2 Appendix B: Levenberg-Marquardt Algorithm

The primary application of the Levenberg-Marquardt algorithm is in the least squares curve fitting problem. Having data points (x_i, y_i) of a set of N measured data the maximum likelihood estimate of the model parameters $\mathbf{a} = \{a_1, ..., a_M\}$ is obtained by minimizing the quantity chi-square χ^2:

$$\chi^2(\mathbf{a}) = \sum_{i=1}^{N} (y_i - f(x_i, \mathbf{a}))^2 \ . \tag{6.13}$$

Here, we omitted for simplicity the individual standard deviation of each measured data point.

Like other numeric minimization algorithms, the Levenberg-Marquardt algorithm is an iterative procedure. To start a minimization, the user has to provide an initial guess for the parameters vector $\mathbf{a} = \{a_1, ..., a_M\}$. In each iteration step, the parameter vector $\mathbf{a}$ is replaced by a new estimate $\mathbf{a} + \boldsymbol{\delta}$. To determine $\boldsymbol{\delta}$, the function $f(x, \mathbf{a} + \boldsymbol{\delta})$ is approximated linearly:

$$f(x, \mathbf{a} + \boldsymbol{\delta}) \approx f(x, \mathbf{a}) + \underline{\underline{\mathbf{J}}} \cdot \boldsymbol{\delta} \tag{6.14}$$

The matrix $\underline{\underline{\mathbf{J}}}$ is the *Jacobian matrix* containing the partial derivatives of the function f according to the parameter a_j:

$$J_{ij} = \frac{\partial f(x_i, \mathbf{a})}{\partial a_j} \tag{6.15}$$

From the first order approximation of $f(x, \mathbf{a} + \boldsymbol{\delta})$ in (6.14) we obtain for χ^2:

$$\chi^2(\mathbf{a}+\boldsymbol{\delta}) = \sum_{i=1}^{N}\left(y_i - f(x_i,\mathbf{a}) - \mathbf{J}_i \cdot \boldsymbol{\delta}\right)^2 \quad (6.16)$$

or in vector notation

$$\chi^2(\mathbf{a}+\boldsymbol{\delta}) \approx \left\| \mathbf{y} - \mathbf{f}(\mathbf{a}) - \underline{\mathbf{J}} \cdot \boldsymbol{\delta} \right\| \quad (6.17)$$

Taking the derivative with respect to $\boldsymbol{\delta}$ and setting the result to zero to find the minimum gives:

$$\left(\underline{\mathbf{J}}^T \cdot \underline{\mathbf{J}}\right) \cdot \boldsymbol{\delta} = \underline{\mathbf{J}}^T \cdot (\mathbf{y} - \mathbf{f}(\mathbf{a})) \quad (6.18)$$

This is the *Gauss-Newton algorithm* to solve a set of linear equations for $\boldsymbol{\delta}$. Levenberg [1] replaced this equation by

$$\left(\underline{\mathbf{J}}^T \cdot \underline{\mathbf{J}} - \lambda \cdot \underline{\mathbf{E}}\right) \cdot \boldsymbol{\delta} = \underline{\mathbf{J}}^T \cdot (\mathbf{y} - \mathbf{f}(\mathbf{a})) \quad (6.19)$$

where $\underline{\mathbf{E}}$ is the identity matrix. The (non-negative) damping factor λ is adjusted at each iteration. If the reduction of χ^2 is rapid, a smaller value of λ can be used, and the algorithm is similar to the Gauss-Newton algorithm. Vice versa, if the iteration is low, λ can be increased and the step will be taken approximately in the direction of the gradient.

Marquardt [2] improved the algorithm by scaling each component of the gradient according to the curvature so that there is larger movement along the directions where the gradient is smaller. This avoids slow convergence in the direction of small gradient. For that purpose Marquardt replaced the identity matrix $\underline{\mathbf{E}}$ with the diagonal matrix consisting of the diagonal elements of $\left(\underline{\mathbf{J}}^T \cdot \underline{\mathbf{J}}\right)$, resulting in the Levenberg–Marquardt algorithm:

$$\left(\underline{\underline{J}}^T \cdot \underline{\underline{J}} + \lambda \cdot \left(\underline{\underline{J}}^T \cdot \underline{\underline{J}}\right)\right) \cdot \underline{\delta} = \underline{\underline{J}}^T \cdot \left(\underline{y} - \underline{f(a)}\right). \tag{6.20}$$

The choice of the damping factor λ is not obvious at all. Marquardt recommended starting with a value $\lambda = \lambda_0$ and a factor $\nu > 1$. Depending on the value of χ^2 the damping factor will be replaced after each step by either λ/ν or $\lambda \cdot \nu$.

The Levenberg-Marquardt algorithm is a very popular curve-fitting algorithm used in many software applications for solving generic curve-fitting problems. However, beware of that also this algorithm finds only a local minimum like all other iterative procedures, not a global minimum.

7 References

[1] Palik, E. D. (Ed.), (1985) *Handbook of Optical Constants of Solids I*, Academic Press, San Diego; and (1991) *Handbook of Optical Constants of Solids II*, Academic Press, San Diego, and (1998) *Handbook of Optical Constants of Solids III*, Academic Press, San Diego.

[2] Bass, M. (Ed.) (1994), *Handbook of Optics, Vol. 2: Devices, Measurements, and Properties*, 2nd ed., McGraw-Hill Professional, New York.

[3] *CRC Handbook of Chemistry and Physics*, CRC Press, Taylor & Francis Group, Boca Raton, 95th edition 2014-2015.

[4] Adacho, S. (1999), *Optical Constants of Crystalline and Amorphous Semiconductors: Numerical Data and Graphical Information*, Kluwer Academic Publications

[5] Ward, L (1994), *The Optical Constants of Bulk Materials and Films*, Institute of Physics Publishing; 2nd edition.

[6] Born, M., Wolf, E. (1975), *Principles of Optics*, Pergamon Press, Oxford.

[7] Stratton, J. A. (1941) *Electrodynamic Theory*, McGraw Hill, New York.

[8] Jackson, J. D. (1981), *Classical Electrodynamics*, 2nd edition, John Wiley & Sons, New York.

[9] Quinten, M. (2011), *Optical Properties of Nanoparticle Systems - Mie and Beyond*, Wiley-VCH, Berlin.

[10] Ashcroft, N. W., Mermin, N. D. (1976), *Solid State Physics*, Saunders College, Philadelphia, International Edition.

[11] Lorentz, H. A. (1895), Versuch einer Theorie der electrischen und optischen Erscheinungen in bewegten Körpern, E. J. Brill, Leiden.

[12] Drude, P (1900), Zur Elektronentheorie der Metalle, Part 1, Ann. Physik, **306**, 566-613.

[13] Drude, P (1900), Zur Elektronentheorie der Metalle, Part 2, Ann. Physik, **308**, 369-402.

[14] J. Leng, J. Opsal, H. Chu, M. Senko, D.E. Aspnes (1998), Analytic representations of the dielectric functions of materials for device and structural modeling, Thin Solid Films **313-314**, 132-136.

[15] Brendel, R., Bormann, D. (1992) An Infrared dielectric function model for amorphous solids. J. Appl. Phys., **71**, 1-6.

[16] Kim, C. C., Garland, J. W., Abad, H., Raccah, P. M. (1992) Modeling the Optical Dielectric Function of Semiconductors: Extension of the Critical-Point Parabolic-Band Approximation. Phys. Rev., **B 45**, 11749-11767.

[17] Tauc, J., Grigorovici, R., and Vancu, A. (1966), Optical properties and electronic structure of amorphous germanium. phys. stat. sol., **15**, 627-637.

[18] Jellison Jr., G. E., Modine, F. A. (1996), Parametrization of the optical functions of amorphous materials in the interband region. Appl. Phys. Lett., **69**, 371-373. Erratum, Appl. Phys. Lett., **69**, 2137 (1996).

[19] A. S. Ferlauto, G. M. Ferreira, J. M. Pearce, C. R. Wronski, R. W. Collins, Xunming Deng, Gautam Ganguly (2002), Analytical model for the optical functions of amorphous semiconductors from the near-infrared to ultraviolet: Applications in thin film photovoltaics. J. Appl.. Phys. **92**, 2424-2436.

[20] O'Leary, S. K., Johnson, S. R., Lim, P. K. (1997), The relationship between the distribution of electronic states, the optical absorption spectrum of an amorphous semiconductor: An empirical analysis. J. Appl. Phys., **82**, 3334-3340.

[21] Urbach, F. (1953), The long wavelength edge of photographic sensitivity and of the electronic absorption of solids, Phys. Rev., **92**, 1324.

[22] Allen, P. B. (1971), Electron-Phonon Effects in the Infrared Properties of Metals, *Phys. Rev.*, **B 3**, 305-320.

[23] Forouhi, A.R., Bloomer, I. (1986), Optical dispersion relations for amorphous semiconductors and amorphous dielectrics, *Phys. Rev.*, **B 34**, 7018-7026.

[24] Kronig, R. (1926), On the theory of the dispersion of X-rays, J. Opt. Soc. Am., **12**, 547-557.

[25] Kramers, H. A. (1927), La diffusion de la lumière par les atomes, *Atti Cong. Intern. Fisica*, (Transactions of Volta Centenary Congress, Como), **2**, 545-557.

[26] D.M. Roessler, Kramers - Kronig analysis of reflection data:, British J. Appl. Phys. **16** (1965) 1119-1123.
[27] D.M. Roessler, Kramers - Kronig analysis of reflectance data: III. Approximations, with reference to sodium iodide, British J. Appl. Phys. **17** (1966) 1313-1317.
[28] P. J. Riu and C. Lapaz, Practical limits of the Kramers-Kronig relationships applied to experimental bioimpedance data, Annals of the New York Academy of Sciences **873**, 374-380 (1999).
[29] Sellmeier, W. (1871), Zur Erklärung der abnormen Farbenfolge im Spectrum einiger Substanzen, Annalen der Physik und Chemie, **219**, 272-282.
[30] Cauchy, A. L. (1830) Sur la réfraction et la réflexion de la lumière, Bulletin de Férussac, tomé **14**, 6-10.
[31] Cauchy, A. L. (1836), *Mémoire sur la Dispersion de la Lumière*, J. G. Calve, Prague.
[32] Conrady, A. E. (1929), *Applied Optics and Optical Design*, Oxford University Press, London.
[33] Conrady, A. E. (1958), *Applied Optics and Optical Design, Part I*, Dover Publications, Inc. NewYork.
[34] Conrady, A. E., and Kingslake, R. (1960), *Applied Optics and Optical Design, Part II*, Dover Publications, Inc., NewYork.
[35] Herzberger, M. (1959), Colour correction in optical systems and a new dispersion formula, Opt. Acta (London), **6**, 197-215.
[36] Herzberger, M., Salzberg, C. D. (1962), Refractive Indices of Infrared Optical Materials and Color Correction of Infrared Lenses, J. Opt. Soc. Am., **52**, 420-424.
[37] Hartmann, J. (1898), Ueber eine einfache Interpolationsformel für das prismatische Spectrum, Publicationen des astrophysikalischen Observatoriums zu Potsdam, **42**, 4-29.
[38] Quinten, M., Kreibig, U. (1993) Absorption and Elastic Scattering of Light by Particle Aggregates. Appl. Opt., **32**, 6173-6182.
[39] Lichtenecker, K. (1926) Die Dielektrizitätskonstante natürlicher und künstlicher Mischkörper. Phys. Z., **27**, 115-158.
[40] Beer, A. (1853) *Einleitung in die höhere Optik*, Vieweg, Braunschweig.

[41] Gladstone, J. H., Dale, T. P. (1863) Researches on the refraction and dispersion and sensitiveness of liquids. Phil. Trans., **153**, 317-343.
[42] Landau, L. D., Lifshitz, E. M. (1974) *Lehrbuch der Theoretischen Physik VIII: Elektrodynamik der Kontinua*, Akademie Verlag, Berlin.
[43] Looyenga, H. (1965) Dielectric constants of heterogeneous mixtures. Physica, **31**, 401-406.
[44] Garnett, J. C. M. (1904) Colours in metal glasses and in metallic films. Phil. Trans. Royal Soc. London, **A 203**, 385-420.
[45] Bruggeman, D. A. G. (1935) Berechnung verschiedener physikalischer Konstanten von heterogenen Substanzen. I. Dielektrizitätskonstanten und Leitfähigkeiten der Mischkörper aus isotropen Substanzen. Ann. Phys. (Leipzig), **24**, 636-679.
[46] Piredda, G., Smith, D. D., Wendling, B., Boyd, R. W. (2008), Nonlinear optical properties of a gold-silica composite with high gold fill fraction and the sign change of its nonlinear absorption coefficient. J. Opt. Soc. Am., **B 25**, 945-950.
[47] Born, M., Wolf, E. (1975), *Principles of Optics*, Pergamon Press, Oxford.
[48] Pedrotti, F., Pedrotti, L., Bausch, W., Schmidt, H. (2005), *Optik für Ingenieure, Grundlagen*, 3rd ed., Springer, Berlin.
[49] Hecht, E. (2001), *Optics*, Pearson Education Limited, Harlow, 4th edition
[50] Bergmann, L., Schäfer, C. (1993) *Lehrbuch der Experimentalphysik, Vol. 3, Optik*, Neunte Auflage, W. de Gruyter, Berlin.
[51] Tien, P. K., Ulrich, R., Martin, R. J. (1969), Modes of propagating light waves in thin deposited semiconductor films, Appl. Phys. Lett., **14**, 291-294.
[52] Harris, J. H., Shubert, R., Polky, J. N. (1970), Beam Coupling to Films, J. Opt. Soc. Am., **60**, 1007-1016.
[53] Tien, P. K., Ulrich, R. (1970), Theory of Prism-Film Coupler and Thin-Film Light Guides, J. Opt. Soc. Am., **60**, 1325-1337.
[54] Ulrich, R., Torge, R. (1973), Measurement of Thin Film Parameters with a Prism Coupler, Appl. Opt., **12**, 2901-2908.

[55] Swanepoel, R. (1983), Determination of the thickness and optical constants of amorphous silicon, J. Phys. E: Sci. Instrum., **16**, 1214-1222.

[56] S. P. F. Humphreys-Owen, Comparison of reflection methods for measuring optical constants without polarimetric analysis, and proposal for new methods based on the Brewster angle, Proc. Phys. Soc. (London) **77**, 949-957 (1961)

[57] J. Strong, *Procedures in Experimental Physics*, 1st ed., Prentice-Hall Inc., New York (1938) 376.

[58] M.A. Archard, *Workshop on Optical Property Measurement Techniques*, Ispra, Italy 27-29 October 1987, Commission of the European Communities, May (1988) 73.

[59] H.E. Bennet and W.F. Koehler, Precision Measurement of Absolute Specular Reflectance with Minimized Systematic Errors, J. Opt. Soc. Am., **50** (1960) 1.

[60] A. Voss, W. Plass and A. Giesen, Simple high-precision method for measuring the specular reflectance of optical components, Appl. Opt., **33** (1994), 8370.

[61] V.R. Weidner and J.J. Hsia, NBS specular reflectometer–spectrophotometer, Appl. Opt., **19** (1980) 1268.

[62] Behrens, H., Ebel, G. (Eds.), (1981) *Physics Data: Optical Properties of Metals, Pt. 1, 2*, Fachinformationszentrum Energie, Physik, Mathematik GmbH, Germany.

[63] Levenberg, K. (1944), A Method for the Solution of Certain Problems in Least Squares, *Quart. Appl. Math.*, **2**, 164-168.

[54] Marquardt, D. (1963), An Algorithm for Least-Squares Estimation of Nonlinear Parameters, SIAM J. Appl. Math., **11**, 431-441.

[65] Press, W. H., Teukolsky, S. A., Vetterling, W. T., and Flannery, B. P. (2002), *Numerical Recipes in C++. The art of scientific computing*, 2nd ed., Cambridge University Press, Cambridge.

[66] Drude, P. (1889), Ueber Oberflaechenschichten, I. Theil, Ann. d. Physik u. Chemie, **36**, 532-560, and II. Theil, 865-897.

[67] Azzam, R. M. A. and Bashara, N. M. (1987), *Ellipsometry and Polarized Light*, 2nd ed., North-Holland, Amsterdam.

[68] Tompkins, H.G., McGahan, W.A. (1999), *Spectroscopic Ellipsometry and Reflectometry*, John Wiley & Sons, Inc., NewYork.
[69] Tompkins, H.G. (1993), *A User's Guide to Ellipsometry*, Academic Press, San Diego, and (2006) Dover Publications, Inc., New York.
[70] Tompkins, H. G., Irene, E. A., Haber, E. A. (2005), *Handbook of Ellipsometry (Materials Science and Process Technology)*, William Andrew Inc., New York.
[71] Tompkins, H. G., Irene, E. A. (Eds.) (2006), *Handbook of Ellipsometry*, Springer, Berlin.
[72] Fujiwara, H. (2007), *Spectroscopic Ellipsometry: Principles and Applications*, 1st ed., John Wiley & Sons, Inc., New York.
[73] Röseler, A. (1990), *Infrared Spectroscopic Ellipsometry*, Akademie-Verlag, Berlin.
[74] Collins, R. W. (1990), Automatic rotating element ellipsometers: calibration, operation, and real-time applications, Rev. Sci. Instrum., **61**, 2029-2062.
[75] Decker, M. M., Mueller, H. (1957), Transmitting Data by Light Modulation, Control Eng., **4**, 63-67.
[76] Billardon, M., Badoz, J. (1966), Birefringence Modulator, C. R. Acad. Sci., **B 262**, 1672-1675.
[77] Kemp, J. C. (1969), Piezo-Optical Birefringence Modulators: New Use for a Long Known Effect, J. Opt. Soc. Am., **59**, 950-954.
[78] Jasperson, S. N., Schnatterly, S. E. (1969), An Improved Method for High Reflectivity Ellipsometry Based on a New Polarization Modulation Technique, Rev. Sci. Instrum., **40**, 761-767.
[79] Jasperson, S. N., Burge, D. K., O'Handley, R. C. (1973), A modulated ellipsometer for studying thin film optical properties and surface dynamics, Surf. Sci., **37**, 548-558.
[80] Jellison Jr., G. E., Modine, F. A. (1990), Two-channel polarization modulation ellipsometer, Appl. Opt., **29**, 959-974.
[81] Jellison Jr., G. E., Modine, F. A. (1997), Two-modulator generalized ellipsometry: experiment and calibration, Appl. Opt., **36**, 8184-8189.
[82] Jellison Jr., G. E., Modine, F. A. (1997), Two-modulator generalized ellipsometry: theory, Appl. Opt., **36**, 8190–8198.

[83] Woollam, J. A., Snyder, P.G. (1992), Variable Angle Spectroscopic Ellipsometry, in *Encyclopedia of Materials Characterization: Surfaces, Interfaces, Thin Films*, (Eds. C. R. Brundle, C. A. Evans, and S. Wilson), Butterworth-Heinemann, Boston, pp. 401-411.

[84] Woollam, J. A., Johs, B., Herzinger, C. M., Hilfiker, J., Synowicki, R., Bungay, C. L. (1999), Overview of Variable Angle Spectroscopic Ellipsometry (VASE), Part I: Basic Theory and Typical Applications, Critical Reviews of Optical Science and Technology, **CR72**, 2-28.

[85] Johs, B., Woollam, J. A., Herzinger, C. M., Hilfiker, J., Synowicki, R., Bungay, C. L. (1999), Overview of Variable Angle Spectroscopic Ellipsometry (VASE), Part II: Advanced Applications, Critical Reviews of Optical Science and Technology, **CR72**, 29-58.

[86] Jellison Jr., G. E. (1993), Data analysis for spectroscopic ellipsometry, Thin Solid Films, **234**, 416-422.

[87] Jellison Jr., G. E. (1998), Spectroscopic ellipsometry data analysis: measured versus calculated quantities, Thin Solid Films, **313–314**, 33-39.

[88] Gerfin, T., Grätzel, M. (1996), Optical properties of tin-doped indium oxide determined by spectroscopic ellipsometry, J. Appl. Phys., **79**, 1722-1729.

[89] Lawrence Berkeley National Lab, Berkeley, California. Data available via http://windows.lbl.gov/materials/chromogenics

[90] Institute of Physical Chemistry, University of Cologne, Germany, ellipsometric measurements on ITO films.

[91] Institute of Physical Chemistry, University of Cologne, Germany, ellipsometric measurements on SuperYellow films.

www.ingramcontent.com/pod-product-compliance
Lightning Source LLC
Chambersburg PA
CBHW050114230526
45470CB00004B/1827